FLOORING

by Derek Phillips FRIBA MCD MArch

A Design Centre Publication

...ustrial Design 1969

Contents

Living room in house at Allestree, Derby.
A relationship between carpet, quarry tile and concrete pavior providing a rich variety of contrast and pattern.
Architects: Diamond Redfern and Partners
Photographer: W. E. Middleton & Son Ltd

Introduction

Whilst my chief qualification for writing about flooring is that, as an architect, I have to make the right choice of floors for my clients, I have in addition a large and growing family who put all floor surfaces to a severe consumer test.

I am told that the women's magazines and national press are constantly receiving correspondence requesting advice on this subject, which appears to be an area where more mistakes are made than in other fields of domestic choice.

Definition of subject matter

Flooring is restricted to the actual surfaces on which you walk; eg the carpet, not the screed or timber boarding on which the carpet rests; the quarry tile, not the site concrete on which the quarry tile is bedded. It is useful to have some information concerning the structural floor (whether this is suspended or solid) because this factor may limit choice of flooring material, and this is covered in the first section, but the main intention of the book is to give some realistic basis for the choice of the floor finish itself.

The criteria on which the choice of flooring should be made is dealt with under the general headings of durability, comfort, appearance, maintenance, laying requirements and cost.

The relative costs of 'laid' floor finishes are quoted as an important factor of choice, accepting the possible danger that the price of labour in laying a material varies from place to place, and more particularly with area of floor—small floors cost more per square yard than larger areas. The only general advice on cost is that, with flooring perhaps more than any other item, you get what you pay for, and economies are generally expensive in the end.

These general criteria are followed by a discussion of the specific requirements of all the main elements of accommodation in a house with illustrations of examples of each type. Clearly there are no 'perfect' solutions; flooring is a matter of taste and of compromise, therefore, there may be many satisfactory solutions to the same problem.

Information about manufacturers' products follows, classified into the different types of available flooring, which are listed alphabetically; having established the criteria which are important for a particular room, this section is designed to aid the final choice.

All manufacturers offer advice on how their materials should be laid, and, once laid, how they should be looked after. It is wise to follow this advice.

History of flooring

Floor finishes are timeless in that the earliest caveman used skins and hides, and we still do today. Water-worn pebbles, so characteristic of our modern courtyard plans, were used in neolithic times, and the Romans used mosaic, cut stone or marble—all flooring materials which are popular today. Likewise, wool carpets are of great antiquity and were used in ancient Egypt and China. They were introduced into England at the time of the Crusades; the manufacture of carpets here started at Axminster in 1755. We may even see a return to the 'rush-covered' floors of early biblical times; rushes were woven into mats in the fifteenth century in England, and these are in common use today. But though natural materials do not change, the greatest growth has been in synthetics.

14th century timber floor, still in use at the Moot Hall, Elston. *Radio Times* Hulton Picture Library

Mosaic floor from Roman villa at Bignor, Sussex. *Radio Times* Hulton Picture Library

Stone flagged floor in continuous use from the 16th century. Late Tudor hall. *Radio Times* Hulton Picture Library

In 1844 Elija Galloway invented Kamptulicon, an amalgam of natural rubber and cork, and by 1860 Frederick Walton patented linoleum. Early linoleum was made in a certain thickness (4.5 mm) known as 'battleship' lino, and in limited colours. Since then it has become available in all colours. The interwar years saw the growth of cheap printed patterned linoleum and this, together with Congoleum, a printed felt based material, was used almost universally as low priced flooring, placed over paper underlays on boarded floors.

A revolution in flooring took place in the years after the war with the manufacture in 1946 in this country of the original asbestos 'thermoplastic tile', the forerunner of the more sophisticated vinyl asbestos and other vinyl types of today. The colours of the first cheap ranges of thermoplastic tiles were poor (black and dark browns), and with a general increase in standards of living, these have been largely replaced by the vinyl asbestos or flexible vinyl sheet or tile floors, except in the field of low cost housing.

Patterned linoleum has been followed by reverse printed PVC and tufted carpet, both post war phenomena. The latter was cheap and initially good looking, but early products failed on durability. Quality has greatly improved with a consequent increase in price. It is quite normal now for the tenants of low rent housing to use tufted carpet in their living room, installing it over the floor finish of thermoplastic or vinyl asbestos tiles already provided—a double floor finish.

The future

There is no doubt that the development of synthetic materials for flooring will continue with improvements in quality, range of colours, surface texture and pattern. There are, however, signs of some reaction to these materials with a desire to return to the beauty of natural materials such as stone, quarry tiles, cork or timber. This manifests itself in attempts made by manufacturers to copy the appearance of cork or timber strip by means of inlaid vinyl strip floors, or the textured vinyl copies of travertone and Provencal tiles which at a superficial glance can fool you. These tricks have an appeal in terms of initial maintenance, but they can never achieve the beauty and lustre of the real thing, a tile polished over the years, the feel of a timber floor. We admire the beauty of york stone floors laid over three centuries ago.

There will be little left to admire of any of our synthetic floors in the centuries ahead. It is in this field of 'quality' of flooring that the synthetic floors never quite 'make it'; they are best when they stick to their own ground, that of achieving good clean interiors with a pleasant range of colours, ease of laying and reasonable maintenance characteristics. For real quality in interiors, with the right 'feel', tread and appearance, in the few interiors in which this appears to matter, there is a good case for using the time honoured natural materials which can improve with age and use.

Construction

The nature of the floor construction is important when making a choice of all types of flooring. The commonest trouble stems from damp rising from the sub-floor and causing deterioration to floor finishes.

There are basically two types of floor construction; suspended floors and solid floors.

Suspended Floors

These will be found in many old houses at ground level, and in all small scale domestic property at other levels; at ground level consisting of timber boards laid on floor joists, supported in turn by timber plates on 'sleeper' walls.

On upper floors, the boards will be supported on floor joists spanning between either cross walls or intermediate structural partition walls.

A characteristic which most affects flooring materials is the tendency of the floor boards to warp and so 'show through' to any thin flooring material such as linoleum or vinyl tiles. This can usually be overcome by using some form of underlay below linoleum, but it will be necessary to place a layer of hardboard or plywood below vinyl, or other forms of thin tile.

A new approach to suspended floors, and one which is likely to become more highly developed with the increased use of industrialised building methods, is the substitution of structural sheet materials such as plywood or chipboard, for the traditional tongued and grooved floor board. This has not only the advantage of speed of erection for the builder, but the additional advantage that thin sheet or tile materials can be stuck or laid directly to it.

Suspended floors will generally transmit heat and sound more easily and will not be 'cold', although due to the necessity to ventilate a suspended ground floor, there is a chance that cold air may be transmitted.

Whilst the problems connected with thin sheet and tile floors can be overcome in the manner suggested, it will not be possible to lay non-resilient floorings—quarry tiles, stone or mosaic—on suspended floors without risking cracking.

Solid Floors

Since the last war the traditional ground level suspended timber floor has given way to concrete sub-floors; these are usually restricted to the ground floors of small scale domestic property, but are used in flats and maisonettes to ensure adequate sound and fire control between levels. Solid floors of various types of construction will be used at all levels in blocks of flats.

They consist, in most cases, of some 6″ of concrete, at the ground level laid on hardcore with a damp proof membrane such as polythene in between.

At upper levels the concrete will be reinforced by steel mesh, or concrete planks will be used to span the required distance. Alternatively reinforced hollow pot floors with concrete poured between are used to lighten the structure.

Solid floors will permit any type of flooring to be used.

If thin tile or sheet floors are chosen, these will be laid on a cement and sand screed. This is necessary because the concrete surface when laid will be too rough for flooring. If hard floors are used, these will be set into a bed of cement mortar which again irons out all the surface

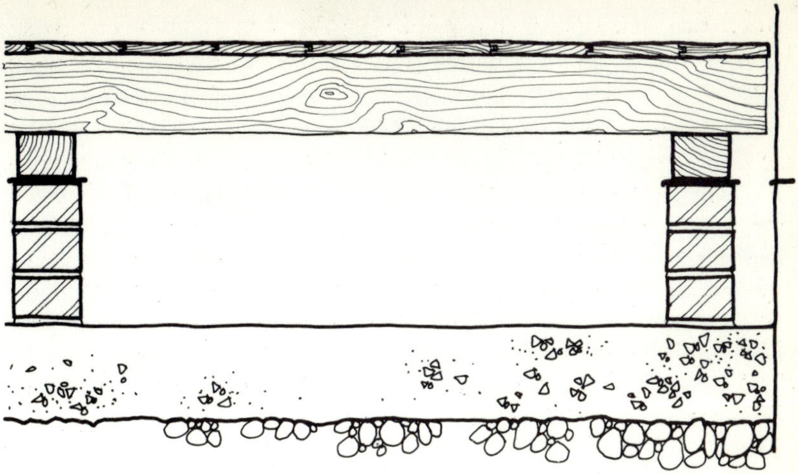

1 Suspended timber floor construction

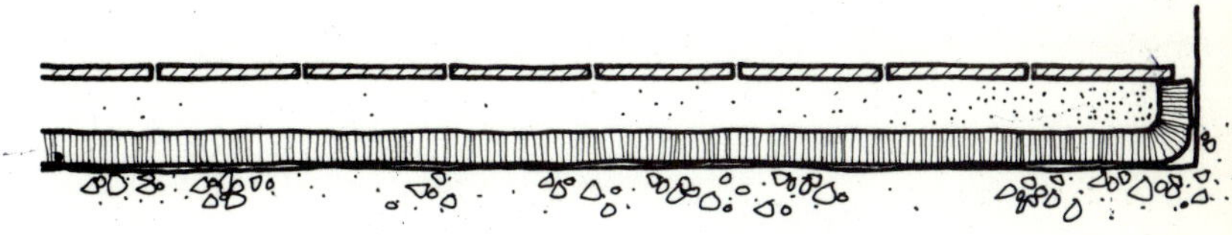

2 Floating floor

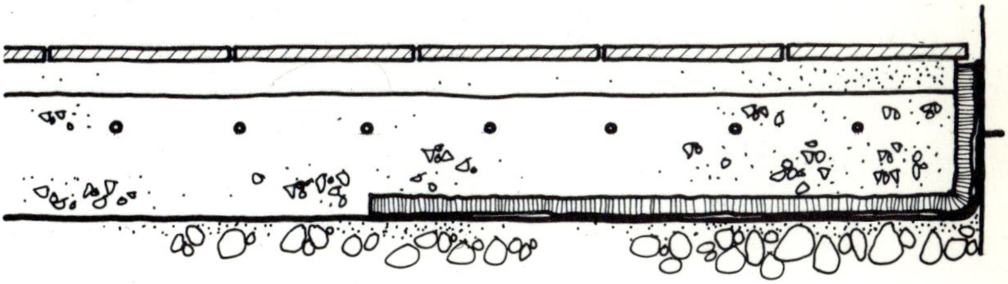

3 Heated floor with edge insulation

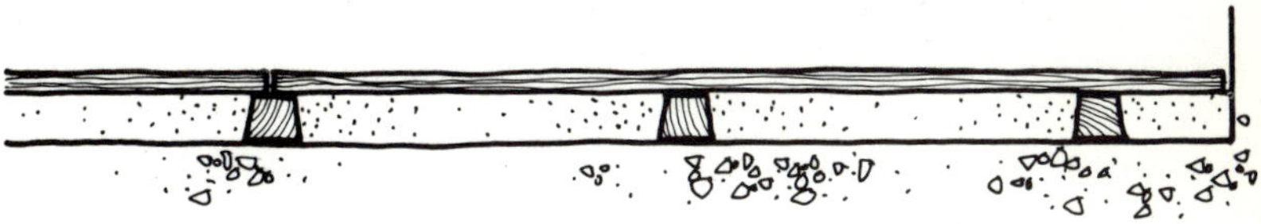

4 Solid floors
a timber on wedge battens set in screed

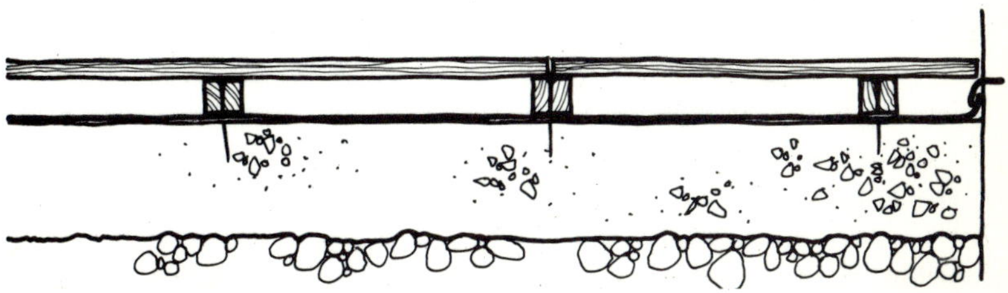

b timber on battens fixed to site concrete

defects of the structural concrete.

If decorative wood floors are used these may be set on an asphalt base, as in the case of wood block or parquet, where an even screed below will be necessary, or in the case of timber strip, special battens may be fixed to the structural floor into which the strip is secretly nailed.

Flooring manufacturers lay great stress on the importance of the screed —a floor is as good as the screed—but this is not necessarily true. It depends upon the type of floor to be used—for example, however good the screed, if a poor floor is placed on it, it still won't be a good floor, It is, however, important for thin stuck tiles or sheet, and anyone who has had the problem of sticking thin tiles to a poorly surfaced screed will agree. Resilient tiles such as flexible vinyl take up the surface eccentricities and look terrible; thermoplastic tiles crack and look even worse.

Damp

Most modern solid floors at ground level will have some form of damp proof membrane consisting of an impervious layer—asphalt, polythene sheet, or bitumen solutions such as the N.C.B.'s 'Synthaprufe', to prevent the movement of rising moisture into the house. On the other hand, in older properties with existing stone or quarry tile floors, such sophistication is lacking and these floors are susceptible to damp; if these floors are left undisturbed they can give years of service, and many beautiful old floors exist which it would be tragic to destroy. If, on the other hand, some form of tile or sheet material is stuck to them, there is a great danger of the floor finish lifting, due to rising damp.

Heated Floors

The most usual form of heated floor (and the cheapest to install) is one in which special electric cables are buried in the screed.

Insulation is generally added at the sides of the house and below the floor to ensure that the minimum heat is lost to the exterior, and clearly in the same way, if insulation in the form of fitted carpeting is placed above, this will impede the transmission of heat into the room; it may also cause the floor to overheat.

The principle of 'off-peak' underfloor heating is to use electricity to heat the floor during the night when electricity rates are lowest, to give off heat during the day, as with other forms of storage heaters; so that they rely on the floor construction to be of sufficient bulk or 'thermal capacity' to go on giving out heat until the next time it can be charged.

Thermoplastic and vinyl asbestos tiles have a comparatively low melting point, approx, 80°F, (26.7°C) and off peak heated floors go up to 90°F (32.2°C) (sometimes as high as 120°F (48.9°C) during the night, particularly if carpet is placed over). Point loads from furniture on such floors cause indentation; so that these finishes are unsuitable.

Manufacturers of most types of flooring claim that underfloor heating can be used, but clearly some floors are more appropriate than others. For example, stone or quarry floors are ideal since they are of 'high thermal capacity' and cannot deteriorate due to higher temperatures. Only the hardest timber floors such as teak are suitable, as floor heating will otherwise cause shrinkage and warping.

Noise

Whilst the type of flooring will affect noise characteristics, hard floors being more noise producing than soft, with carpet way out ahead in quietness, noise should also be considered as an aspect of structure. Solid floors will be more noise resistent than suspended floors and some types of solid floor will be better than others, generally the heavier the better. Alternatively, impact noise can be greatly reduced by means of a 'floating floor' consisting of using a screed laid on top of a fibreglass or other form of quilt, truly isolating it from the sub-floor, walls and skirting.

General Criteria

The importance which should be attached to any particular criteria will depend upon individual circumstances. For example, cost may be of vital importance and it may be necessary for financial reasons to sacrifice durability or appearance in order to meet a specific budget, but it should be quite clear when this choice is made that it will be necessary to renew the floor finish after a limited period. Similarly in areas of very heavy use and wear, as for example in a utility room which children use as a garden access in muddy boots, it is clearly sensible to stress durability at the expense of comfort in terms of warmth or noise. The more use and wear a room has, the fewer options there are—it may be comfortable and the height of luxury to have a carpet in the bathroom, but perhaps wiser to wait until one's children have grown out of the water splashing stage.

The criteria of choice of flooring is discussed under the following headings:

Durability—resistance to wear and other factors
Comfort—warmth, noise, slipperiness, tread or feel
Appearance
Maintenance
Laying Requirements
Costs

Durability

This is often considered to be the most important criteria of all; and clearly in areas of very heavy wear it is of overriding importance. Hard floors on the whole are more durable than soft, but durability is also related to resilience, and whilst rubber is not a hard floor it is extremely durable.

Durability means the length of time that selected qualities (appearance,

Comparative Durability

		Resistance to wear	Resistance to water	Resistance to indentation
Asphalt	Mastic Asphalt	VG	VG	F-P
	Pitch Mastic	G	VG	F-P
Carpets	Axminster Wilton Tufted Cords others	G-P	F-P	F
Ceramics	Clay floor tiles Quarries Mosaic (several types including glass) Bricks	VG-G	VG	VG
Concrete	Granolithic/ jointless	VG-F	VG	VG
	Concrete tiles	VG-G	VG	VG
	Terrazzo/ jointless tiles	VG-F	VG	VG
Cork	Cork tiles	G	P	P
	Vinyl cork	G	VG-F	P
Linoleum	Jute/hessian backing sheet or tile Saturated paper backed tile	G	F	G-P
	Cork carpet	G	P	P
Magnesium Oxychloride		G	P	VG-G
Plastics	Thermoplastic (including vinylised thermoplastic tile	F-P	G	P
	Vinyl asbestos tile	G	G-F	F
	Flexible PVC sheet/tile/ textured or embossed PVC with hessian, cork, felt or cellular backing	VG-F	VG-F	G-F
	Reverse printed PVC	F	F	F
	PVA emulsion	F	P	F
	Polyurethane resin	G	VG	VG
Rubber	Solid rubber/ natural or synthetic/ inlaid Sponge-backed Matting Ribbed Rubber cork sheet	VG-F	G	G-F
Stone	Granite Sandstone (eg York Stone) Marble Quartzite Slate	VG-P	G	VG

Timber	Softwood board/ block/strip	F	P	G-F
	Hardwood board/ block/strip	VG-F	P	VG-F
	Block parquet	VG-F	P	VG-F
	Block mosaic	VG-F	P	VG-F
	Block endgrain	VG-F	P	VG-F
	Plywood/parquet	P	P	G-F
	Chipboard	F	P	VG-F
Composition block		G	G	VG

VG = very good G = good
F = fair P = poor

colour, surface, pile) will last. Therefore, durability in some materials cannot always be equated with wear resistance. For example, in the field of carpets, the synthetic fibres such as nylon are extremely hard wearing, but their appearance after a short time is less good than wool mixtures which are more springy and retain their appearance over very long periods of time, and it is clear that the durability of floors such as carpet or cork should be judged from a different standpoint from hard floors such as stone or concrete. It is important to assess the relevant factors in each case, since durability is affected by many factors not all of which must be taken into account in every case. For example:

Resistance to chipping and cracking
Resistance to scratching and abrasion
Resistance to indentation
Resistance to heat, burning coals, etc
Resistance to acids
Resistance to water and rot
Resistance to oils and fats, alkalis
Resistance to insects—moths, beetles, etc

In a kitchen all these factors must be taken into account, and one would expect to find surfaces which can resist all these attacks. Yet it is surprising how in many kitchens, for reasons of economy, thermoplastic tiles have been used, which do not withstand heat or acids, are subject to cracking, and are impaired by many ordinary household fluids such as paraffin.

If one were to consider durability alone, the best kitchen floor solution would be found amongst the hard floors such as quarry tiles or stone. There are many areas in a house where durability is not of paramount importance—particularly in bedrooms, and here durability will not be given high priority.

An important factor in the durability of a floor is that it should be laid properly, and second, that it should be maintained correctly. If floors are left without proper maintenance, extra abrasion will occur and the life and appearance of the floor will be reduced.

Comfort

This is a subject where subjective evaluation makes a strict analysis difficult. To illustrate this one need only take the question of warmth. A carpet is usually thought to give an impression of warmth, and stone or concrete floor one of cold. But whereas it is not sensible to use fitted carpets with a heated floor, as this will tend to insulate and therefore decrease the efficiency of the heating method, a stone or concrete tile floor is ideal for underfloor heating. From the point of view of temperature, therefore, the 'cold' floor may be hot and the warm floor cold. Which, therefore, is the more comfortable?

Important aspects which contribute to our impression of comfort are as follows:

- Warmth
- Quietness or noise
- Slipperiness
- Feel or tread.

Warmth

The 'warmth' of a floor depends on a number of factors which can be measured, and from these factors a 'coldness' scale expressed in arbitary units can be drawn, as follows:

Marble	20
Concrete Tiles	12
Linoleum	8
Wood	5
Cork	0.7

But if the temperature of the floor is also taken into account a basically 'cold' floor can be made as warm as a basically warm floor, eg a concrete floor at 70°F (21.1°C) feels as warm as a wooden floor at 64°F (17.8°C) to the bare foot.

The answer, therefore, as to whether a floor will feel warm or cold does not lie with the material alone, and the heated concrete tiles in my own living room give a pleasantly warm sensation in winter and a delightfully cool one in summer–whilst the cork floor in our unheated bedroom feels cold in winter, and it has been necessary to add a fur rug to achieve a degree of comfort.

Whether a floor feels cold or warm will depend also on whether the air close to the floor is warm and, of course, on whether the person is wearing shoes or not. No discomfort will be experienced by a person wearing shoes on a concrete floor if the air temperature near the floor is 65°F (18.3°C), whilst discomfort will be experienced on a cork floor if the room has a low temperature close to the floor. So it can be seen that the 'warmth' of a floor is not a quality of the flooring alone.*

*Research by Dr Chrenko.

Warmth

Scale of warmth of flooring

Very good	Carpet Cork Timber	 Softwood/board, block, strip Chipboard
Good	Linoleum Plastic Timber Composition block	 Flexible PVC sheet/tile Hardwood/board, block strip, Plywood parquet
Fair	Asphalt Magnesium Oxychloride Plastic	 Thermoplastic tiles, Vinyl asbestos tiles PVA emulsion Polyurethane resin
Very poor	Ceramics Concrete Stone	

Noise

The problem of noise in domestic interiors is one which will be affected by the nature of the flooring but the amount of noise generated by normal foot traffic on the most noisy types of floor such as clay tiles or concrete, will not generally prove unacceptable. The quietest floors are cork and cork carpet, flexible PVC with various backings, rubber, and, of course, carpeting, the latter particularly having good properties of sound deadening. There are three aspects of noise related to flooring which should be considered:

1. Noise from impact; associated with the floor material itself and the type of shoe sole—leather composition, etc.
2. Noise transmission from the first to ground, or from upper to lower levels in flats; associated with the type of floor construction.
3. Acoustic absorption of room generated sound; it is a well known fact that an empty house reverberates and sounds carry, whereas once the furniture and floor finishes are down, these furnishings absorb sound and the whole house is quiet, and noise does not carry from space to space.

Flooring has an effect on all these aspects. In the first case because hard floors create more noise on impact, in the second the nature of the construction is of prime importance and it may be necessary to provide for 'floating' floor construction where sound insulation between two living units is critical. Finally, in the third case, the best acoustic absorption will be derived from carpets, but cork or cork backed vinyl rubber, etc, have good absorption characteristics also. Hard floors such as thermoplastic tiles on a concrete sub-floor, pitch mastic or magnesium oxychloride, are particularly poor in this respect.

Quietness/impact sound

	Scale of quietness/impact sound of flooring
Very good	Carpet Cork Plastic/Flexible PVC Rubber
Good	Linoleum
Fair	Asphalt Magnesium Oxychloride Plastic/ Thermoplastic tiles – vinyl asbestos tiles PVA emulsion – polyurethane resin Timber/Softwood block/strip Hardwood block/strip Composition block
Poor	Timber/Softwood board } on joists Hardwood board } on joists Plywood parquet Chipboard
Very poor	Ceramics Concrete Stone

Slipperiness

A slippery floor is a real hazard and accounts for a large number of accidents in the home. Nearly 5,000 people in Great Britain die annually from falls (over 50% of total accidental deaths) and of these about half are falls on a single level by slipping or stumbling, of which the vast majority are old people over the age of 65. The slipperiness of a floor can be a matter of life or death.

The coefficient of friction between the sole of a shoe and the floor surface is reduced if the floor is wet in the case of hard surfaces such

Slip

	Scale of slipperiness/dry conditions
Very good	Carpet Cork Rubber Timber/softwood board, block, strip
Good	Asphalt Linoleum Magnesium Oxychloride Plastic/Thermoplastic tiles Vinyl asbestos tiles Flexible PVC/backings PVA emulsion Polyurethane resin Timber/Softwood Hardwood Chipboard Composition block
Fair	Ceramics Concrete Stone Note: These surfaces can be obtained with non-slip finishes

as clay tiles, concrete, or the very smooth surfaces such as vinyl sheet or rubber, and dangerous areas can occur at entrances when people come in from the rain, or in potentially wet areas such as kitchens or bathrooms.

The overzealous polishing of timber boards or block, particularly hardwood, can cause danger, and terrazzo floors should never be polished for this reason. Too much polish on thermoplastic or vinyl asbestos floors can be equally disastrous. Carpets, as with criteria for noise and warmth, come out best, but even here in the case of small areas of carpet or rugs, some form of non-slip underlay should be used over potentially slippery floors, and the area below carpets should be left unpolished.

Tread

The tread or 'feel' of a floor is not easy to define, and most standard works on flooring do not attempt to do so. Yet it is easy to differentiate broad differences between the resilience of hard and soft floors.

There is little evidence to support the idea that hard floors are more 'tiring' to walk on than soft and considerable evidence that walking on deep carpet all day, as do the waiters in plushy restaurants, is deleterious to the feet. What is likely to prove a satisfactory solution is some change of floor finish in different areas of a house, so that a housewife who may spend the majority of her day in the home, will experience a degree of variety of tread.

Resilience

	Scale of Resilience
High	Carpet Cork Linoleum (thick grades) Plastic/Flexible PVC with backings Rubber (very high for sponge-backed types)
Moderate	Linoleum (thin grades) Plastic/Vinyl asbestos PVA emulsion Polyurethane resin Timber
Slight	Asphalt Plastic/Thermoplastic tiles
None	Ceramics Concrete Magnesium Oxychloride

Appearance

Appearance is important to all, but for some people it will be an overriding consideration.

The range of choice of appearance is very great indeed, and in order to simplify it, appearance is broken into a number of categories:

1 Component shape:

a jointless

b sheet materials

c tiles
d strip or board
e block.

There is considerable overlap from one shape to another with different materials, for example PVC flooring is obtainable in tiles, in sheet, and when the sheet has welded seams, it can appear to be jointless. The issue is further confused by the desire on the part of manufacturers to be all things to all men, and PVC is now made with a wood grain finish to look like a timber strip floor, and PVC tiles are made with deeply embossed finishes to resemble marble or even Provencal tiles. In addition to this first breakdown related to size and shape of the flooring component are the following:

2 Surface pattern as superimposed on the component pattern.
3 Surface texture.
4 Colour.

2 Surface pattern

This may be a superimposed pattern of a decorative type, available in many ranges of ceramic tiles, or a simple marbled pattern obtained by mixing colours together. A very wide range of surface pattern is available in certain types of flooring, eg:

Ceramic tiles
PVC sheet and tile
Linoleum
Terrazzo
Carpet

whilst with many types of flooring the surface pattern is limited to flecking and marbling, as in

Rubber
Polyurethane resins
Thermoplastic tiles

Alternatively, most natural materials have an intrinsic surface pattern associated with the material itself—the grain of timber or cork.

Surface pattern is clearly very much a matter of personal taste, but there are the hard facts of maintenance to be taken into account, and whilst architects and others generally prefer to see light plain colours on floors in the PVC or vinyl asbestos ranges, some compromise is generally necessary. Modern ranges of tiles overcome to a great extent the old criticisms of marbled surfaces, by providing light flecking which does not look objectionable, and at the same time does not show the scuff marks and dirt too much.

3 Surface texture

By this I mean the roughness or smoothness of the surface, irrespective of whether there is any superimposed surface pattern. For example, there will be a difference of surface texture between a natural timber floor and an imitation timber strip vinyl floor, although the 'superimposed pattern' of the grain may look alike.

Generally speaking, the rougher the texture the more difficult the material will be to clean, and whilst the vinyl material would be a good solution to a kitchen, a timber floor would have serious disadvantages.

A number of materials have been developed with coarse textures, of dots, lines, triangles and the like, designed to improve the non-slip properties of the floor (eg the ribbed rubber floors), but these have had no widespread use in domestic buildings.

In the field of flexible PVC flooring there are a number of examples of applied surface textures designed to give the flooring the appearance of natural materials such as travertine marble or pebbles, and very persuasive imitations have been produced—they even create the same difficulties of cleaning in the joints and fissures associated with the natural materials.

Some of the most elegant textures are found in the field of ceramics, the vitreous and semi-vitreous tiles.

The effect of floor pattern on appearance
1 Plain colour or texture
2 Traditional clay tile pattern
3 Square tile form—contrast
4 Strip pattern

Overleaf: 'The element of choice appears to be limitless'

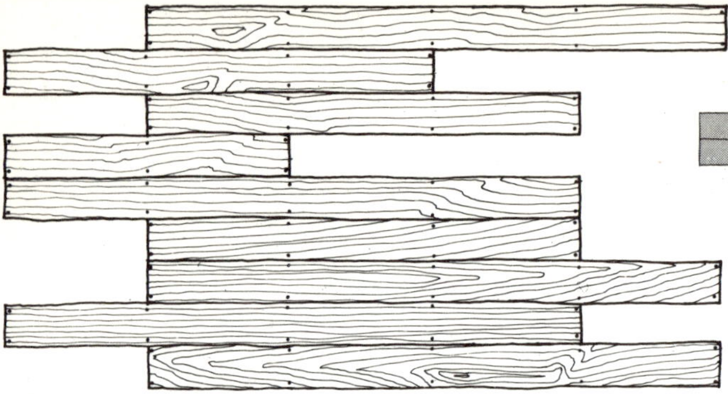

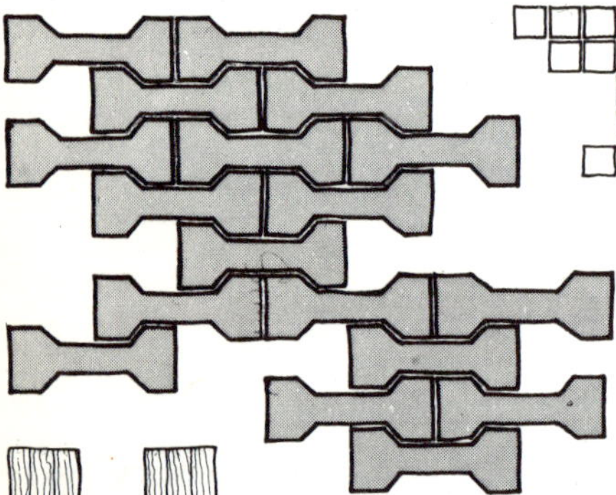
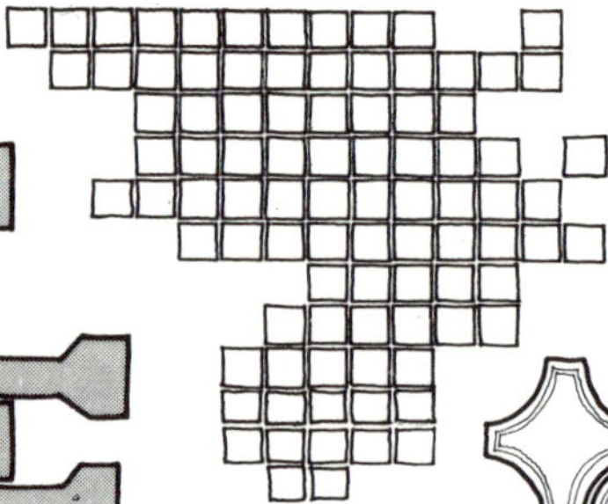
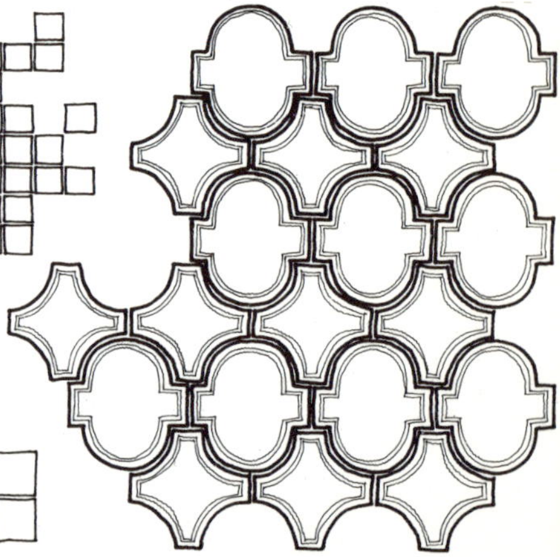
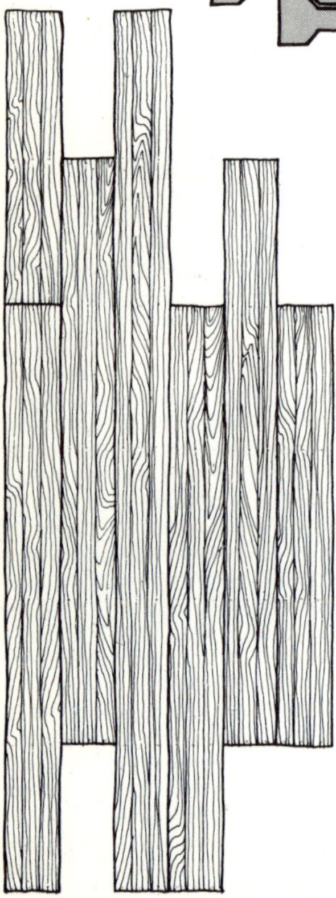
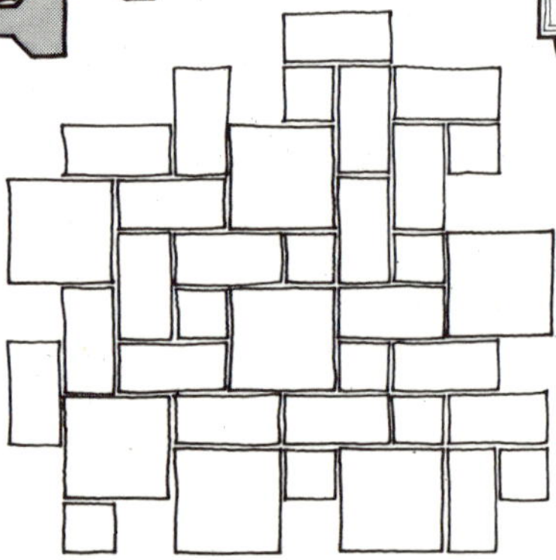
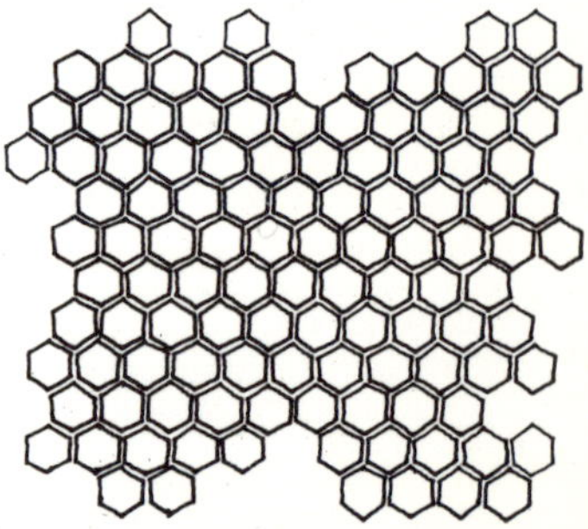

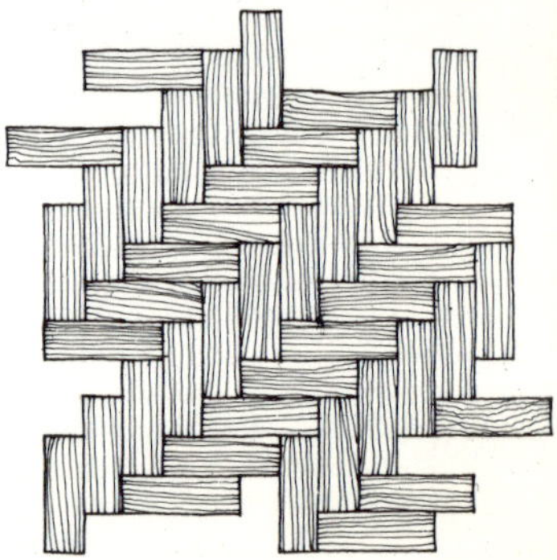

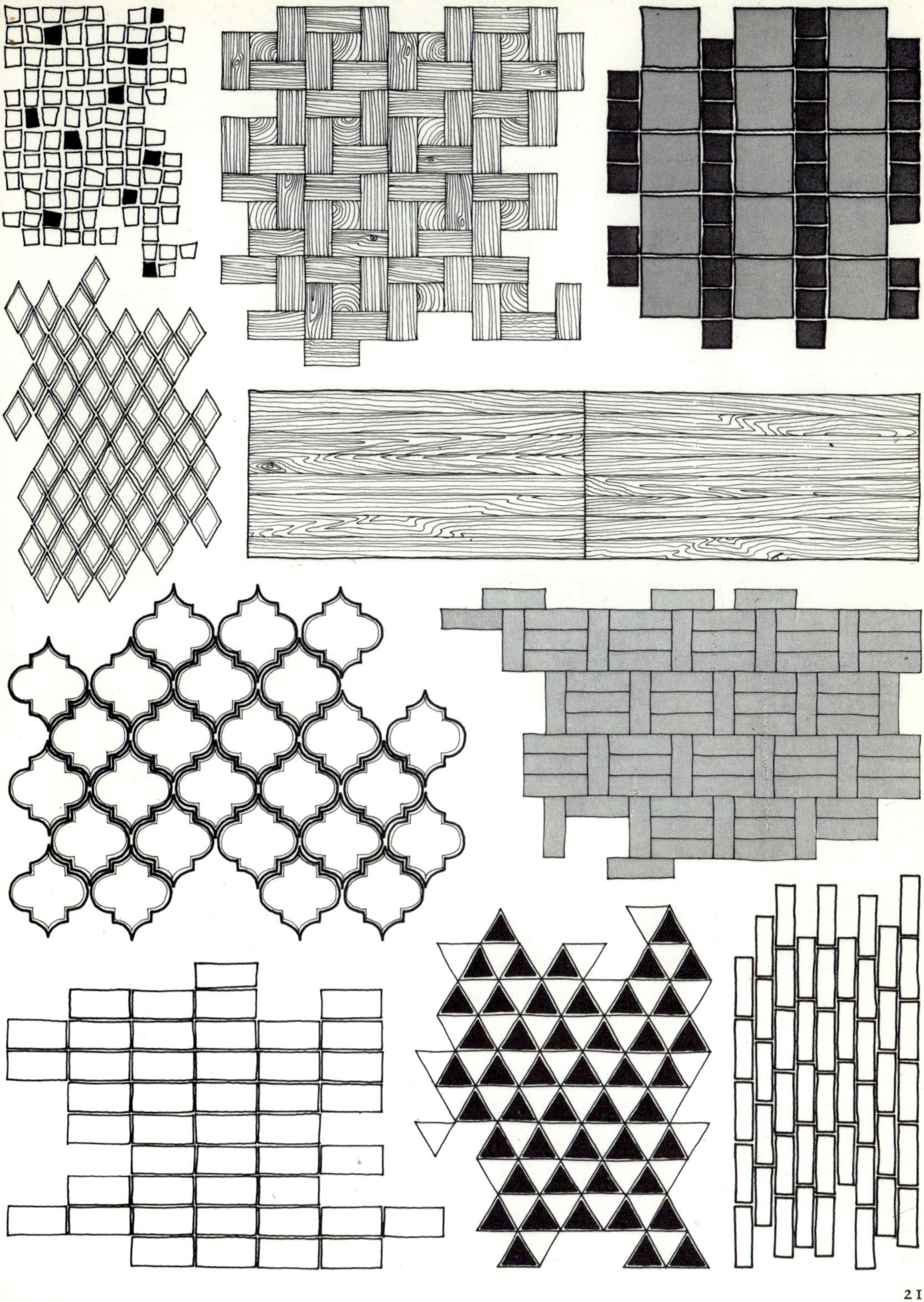

4 Colour

The range of colours available depends upon the type of flooring: in some cases an almost unlimited range whilst in others very limited, due either to the range of natural materials or to the process of manufacture.

1 Limited colour ranges

Asphalt	Black and dark colours
Ceramic/Bricks	Natural blues, browns or reds
Concrete	Restricted range
Cork	Natural cork colour light/dark, few additional colours from dyes
Magnesium Oxychloride	Pink, reds, buffs/plain or marbles
Plastic/PVC Emulsion	Red, buff, green, blue
Plastic/Polyurethane Resin	Red, green, grey or buffed natural
Stone	Variety of natural materials
Timber/Softwood Hardwood	Variety of natural materials
Timber/Chipboard	Light wood colour (darkens when sealed)
Composition block	Light or dark wood with few additional colours from dyes

2 Multi-coloured range

Carpets
Ceramics
Linoleum
Plastic/Thermoplastic
Vinyl Asbestos,
Flexible PVC
Rubber

A great variety of colours is available in all these materials, and new ranges of colours are frequently introduced to the market.

General

When one takes into account the variety of flooring available, of different component size and shape, material and appearance, the element of choice appears to be limitless, and this is often daunting to the selector, but whilst at first sight the choice appears wide open, it is only when the use of a space is fully analysed that many of the options are eliminated as unsuitable and the choice reduced to manageable proportions.

Maintenance

The maintenance of a floor finish is of the utmost importance to its continued durability and good appearance.

Manufacturers all publish detailed information on maintenance of their

General Maintenance

Asphalt	Mastic Asphalt Pitch Mastic	Sweep, wash, emulsion polish
Carpets	Axminster Wiltons Tufted Cords others	Brush, vacuum clean occasional shampoo
Ceramics	Clay floor tiles Quarries Mosaic (several types including glass) Bricks	Sweep, wash, polish (scour with household cleaning powders if very dirty. Quarries: linseed oil initially)
Concrete	Granolithic Jointless Concrete tiles Terrazzo/jointless tiles	Seal, sweep, wash " " " Sweep, wash (do not polish)
Cork	Cork tiles Vinyl cork	Seal, mop, emulsion polish Mop, emulsion polish
Linoleum	Jute/hessian backing sheet or tiles Saturated paper backed tile Cork carpet	Sweep, wash, mop, polish or seal
Magnesium Oxychloride		Initially: linseed oil. Sweep, wash, polish or seal
Plastic	Thermoplastic (incl. vinylised T) tile	Sweep, mop, polish
	Vinyl asbestos tile Flexible sheet/tile/ textured or embossed PVC with hessian backing cork backing felt backing cellular backing Reverse printed PVC PVA emulsion	Sweep, wash, polish
	Polyurethane resin	Sweep and mop
Rubber	Solid Rubber/natural or synthetic/inlaid Sponge-backed Matting Ribbed Rubber cork sheet	Sweep, wash, maybe polish
Stone	Granite Sandstone (eg York Stone) Marble Quartzite Slate	Sweep, wash, scrub
Timber	Softwood board/block/ strip Hardwood board/block/ strip Block parquet Block mosaic Plywood/parquet Chipboard	Sweep, seal, polish and mop
Composition block		Sweep, seal, polish and mop

floors, and it is wise before making a final selection to study their literature to see if its care is within one's own capability, or if it is so time consuming as to result in the floor being left to take care of itself, which in most cases it will not. The two main methods of treating a floor finish after it has been cleaned are by seal or by polish.

1 Seal

Seals are used on cork, linoleum or concrete, but should not be used on thermoplastic tiles, vinyl asbestos or PVC floors.

Seals are of three types:

- a oleo resinous
- b 1 can plastic
- c 2 can plastic

The two can plastic seal uses a hardener separate from the resin which causes setting only when amalgamated. The resin will vary according to the type of use.

2 Polish

The simplest and time honoured method of protecting a floor finish. It consists of applying a protective film which can be removed and replaced when it gets dirty. Modern polishes are either spirit based waxes or water emulsion polishes.

The spirit based waxes can be used on timber, cork, linoleum, concrete or clay tiles, but as all of these waxes contain solvents they cannot be used on such materials as asphalt, rubber and thermoplastic tiles. Emulsion polishes can be applied by hand but they have been developed for machine operation and can be used with all types of floors with the exception of those such as timber, cork or linoleum which might be affected by the water in the emulsion, unless already sealed.

Laying Requirements

This is not intended to be a Do-it-Yourself handbook, and therefore detailed laying instructions for different types of floor are not included, but some of the more general aspects should be stressed.

1 Damp

With one or two exceptions, all floor finishes should be free of damp. This must be solved with the structure of the house, and whilst in new properties there should be no problem, difficulties do arise when laying finishes on the ground floor of old houses where no horizontal Damp Proof Membrane exists.

Asphalt, since it acts as a barrier to damp, is really the only floor which can be laid without a Damp Proof Membrane, but it is not unusual to find concrete or clay tiles which are themselves stable under damp conditions laid without one. It should be noted, however, that under the new Building Regulations a DPM is now a requirement.

2 Screeds

In the case of solid floors, these will comprise a rough surfaced concrete slab varying from 4″ to 6″ in thickness. When quarry, clay, or concrete tiles are used, these will generally be set direct on to this base slab on a cement mortar bed allowing for levelling, but in the case of thin sheet or tile floors, a screed is required. Screeds consist of 1½″ to 2½″ of fine sand and cement, and it is upon the stability and flatness of this surface that the ultimate success of the finished floor will depend—if the screed cracks or moves, so the finished floor will be damaged—if the screed has not had a sufficient 'drying out' period, so the adhesive binding the finished surface will fail. Expert advice should be sought on drying out periods for different screed thicknesses, and a measuring device is available to check moisture content if in doubt.

Some proprietary screeds (Synthanite) have shorter drying out times than cement screed, these can be thinner (1″) and can be laid to a very flat surface.

3 Underfloor Heating

If underfloor methods of heating have been adopted, this will limit the type of floor finish which can be used.

Suitable floorings include the following:

- Ceramics
- Concrete tiles
- Terrazzo tiles
- Cork
- Linoleum
- Plastic/Thermoplastic tiles*
  - Vinyl asbestos tiles*
  - Flexible PVC
- Stone
- Timber/Chipboard
- Composition block

Materials which are laid in jointless form are less suitable than those in tile form, and specialist advice should always be sought.

Selected timber hardwood flooring such as teak can be used, but this must be carried out by a specialist contractor. Fitted carpets will insulate the floor and these are really unsuitable. Rush mats will dry out and deteriorate.

4 Obsolete patterns

Rather as the wall paper industry regularly brings out new ranges to the infuriation of those who just want a couple more rolls of last year's pattern, so there is a tendency for manufacturers of the more popular types of flooring, such as vinyl asbestos or flexible PVC, to change their colours every few years. For a variety of reasons it may be necessary to renew a small area of flooring, and in order to overcome any possible difficulty with supply, it is wise to order 5% extra tiles. It is less easy to repair sheet.

*upper limit of floor temperature should be designed not to exceed 80°F (26.7°C).

Cost

The costs of floor finishes vary enormously from 'laid' prices of 11/– per square yard for the cheapest thermoplastic tile to 285/– per sq. yd. for Westmorland Slate, and for the very exotic marbles, twice as much again. The identical flooring will vary in cost due both to the size of contract (very small areas are relatively more expensive), and the location. Building costs, and therefore the laying costs, are more expensive in some parts of the country.

One lesson above all other I have learnt concerning flooring; it is that the flooring which costs the least initially is not necessarily the most economic. The cheapest carpets are often the worst value of all—our first choice barely lasted a year, yet we are still using a patterned Axminster carpet inherited from a relative, originally used as a stair-carpet and now sewn up to fit a bedroom, which has been in continuous use for 25 years and shows little sign of wear. The thermoplastic tiles in our kitchen started to deteriorate within a short time, and when the opportunity came after about five years to redesign our kitchen, we replaced them with a riven slate floor. This has been down for about two years and improves with time—it has an 'indefinite' life. The slate cost about 45/– a yard, supplied only, whilst the thermoplastic tiles cost about 8/–, but despite this, the slate will end up by being a cheaper floor.

Discussion of Individual Rooms

This section deals more specifically with the requirements of different rooms in a house. The more sophisticated areas of larger houses such as libraries, breakfast rooms, saunas, are not mentioned.

Ground floor
- Living room
- Dining room
- Kitchen
- Playroom

Upper floor
- Bedroom
- Bathroom and WC

Circulation areas
- Corridor, halls
- Stairs

Exterior Spaces
- Garages
- Patios and terraces

General note on limitation of choice

The ground floor may be suspended timber construction but more often is a solid concrete floor with a screed. Any type of floor finish can be used on solid construction if it is decided at the planning stage

Comparative Flooring Costs Laid Per Square Yard

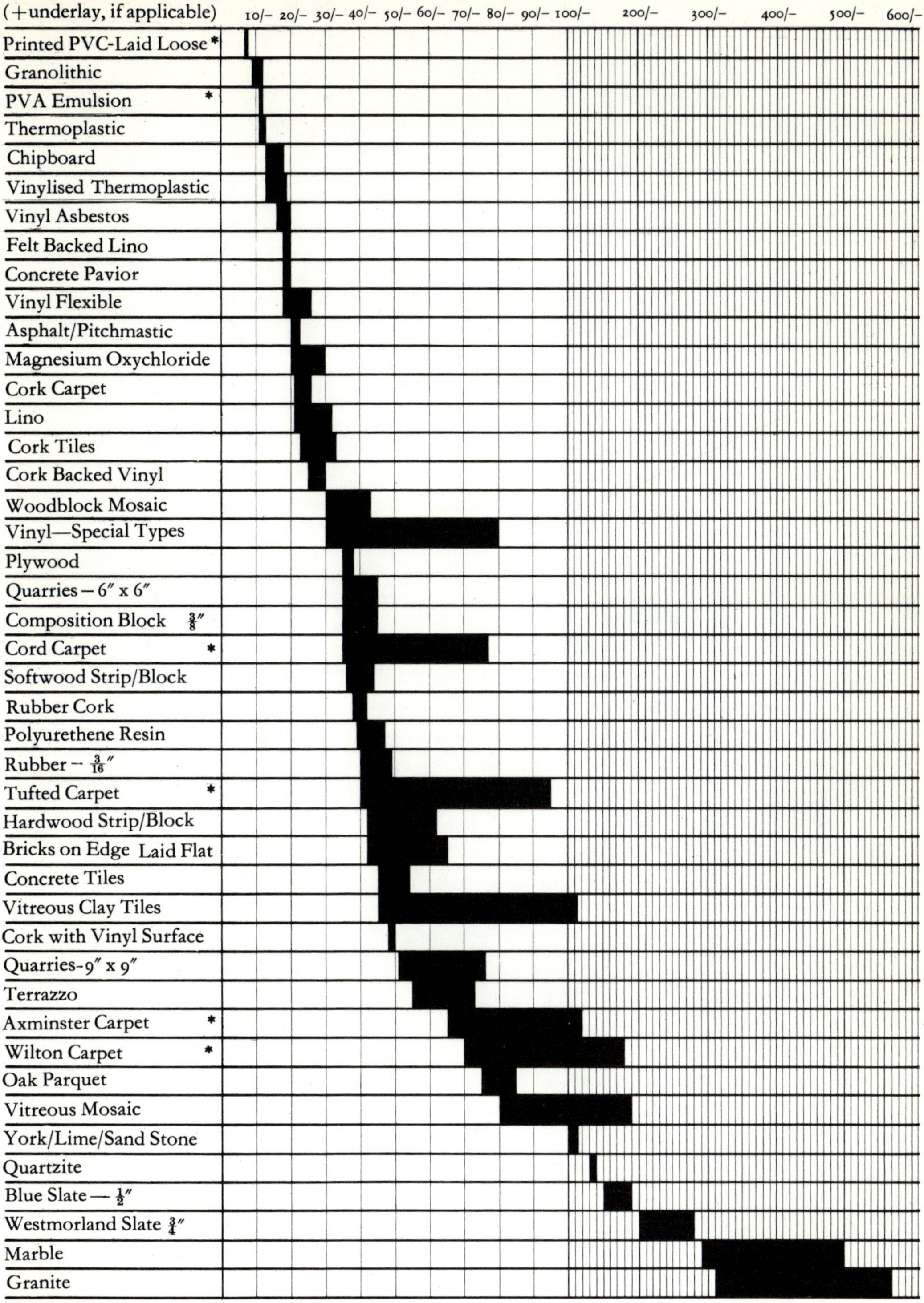

*Excludes laying

of the building, and therefore the choice of finish is not limited. With suspended timber floors at upper (or lower level), the choice is more limited, so that in bedrooms and bathrooms in the average house special constructional techniques would have to be employed if it were decided to use a hard floor such as ceramic tile. Similarly, if thin tile or sheet materials are used the underfloor must have a completely flat surface obtained from using a chipboard or plywood construction, or by adding hardboard or plywood to a boarded floor. These facts must be borne in mind when considering the choice of floor finish.

Opposite: Oak boards with areas of carpet at Balniel House, Colinsburgh in Scotland. Architect: Trevor Dannatt Photographer: John Dewar

Living Rooms

It is impossible to generalise about living rooms; it depends on who is doing the living. If it is a millionaire bachelor, a white carpet might be an elegant solution, if it is a room used continually by a family with children, there are more suitable choices.

What are the requirements? It must be durable, taking wear in its stride, and look good over as long a period of time as one can afford. It must be resistant to those things likely to be spilt—drinks, coffee, ash—and those other things likely to be brought in on the soles of shoes—dirt, mud and grease. It must give a sense of comfort by warmth, or resilience, and should be neither slippery nor noisy.

Combined living and dining room in a house at Hampstead. Linoleum tiles over a heated floor. Architect: Michael Brawne Photographer: Behr Photography

No floor is equally suitable for all these purposes, and there is certainly a case for having a hard surface over areas used principally for traffic in 'through-circulation' rooms with a carpeted area for sitting though some people may prefer fitted carpets throughout. A traditional floor, which, in this country is associated with luxury, is hardwood block or mosaic, or perhaps its more modern equivalent of hardwood strip. This gives a warm rich appearance, particularly if associated with carpet or rugs. When this can be planned, the carpet should be recessed to enable both floors to end up level. Loose rugs can look well but are a source of danger from tripping or slipping.

With open plan living-dining areas it is important that the floor should 'run through' as a visual link between areas which may have little else to unify them. It is often pointed out that single flooring solutions, eg overall carpeting, make for ease of maintenance by a single cleaning method—the vacuum cleaner—true, but this is only one consideration and there are many others. It doesn't necessarily follow that a solution based on easy maintenance is the best solution to all aspects of living. It is fair to say that a good quality carpet answers the needs of a living room best whether this is taken over the whole or part area. What is important is that it must be 'good quality' and capable of withstanding wear and dirt. This is the one area of the house in which economies should not be made—a bedroom carpet which gets little wear can be relatively inexpensive, but the living room carpet should be the best you can possibly afford, if you are not to regret it later (as we have, not once but twice).

A good compromise is a wool carpet with 20% of nylon, either Wilton or Axminster.

Quarry tiles used with a Rya rug to add richness in a house at Shipton-under-Wychwood. Architects: Stout and Litchfield Photographer: Richard Einzig

If carpet is decided upon right from the outset with a new house, this can be a saving in construction as it will be unnecessary to take any other expensive finish over this area—all that will be required is a flat cement and sand screed over the base concrete, into which should be set timber tacking strips for fixing the carpet at the edges. On suspended timber floors, a chipboard underfloor will be adequate for laying carpet. There is one point to bear in mind—that the next owner of the house may not want carpets and the initial savings in construction may be reflected in the sale price.

White mosaic used as a unifying factor throughout all areas of Ferrum House, Harpenden.
Architect: Jack Bonnington
Photographer: Henk Snoek

Dining Rooms

In the small family house, the dining room is often a part of the living room or even an alcove off the kitchen—in these circumstances the floor finish will be governed by the major space, but in the separate dining room used both for family eating or for formal dinner parties there is sometimes a case for making a change from whatever flooring is selected in other living areas.

Dining room floors have broadly speaking the same requirements as living rooms, being subject to the same factors, but there will generally be less wear.

Provided that the room is well heated, hard floors are ideal as crumbs and spilt food can be brushed up easily. Carpets are less successful in this respect and are apt to show signs of staining in particular areas of the room related to food serving and children's seats.

Hardwood floors are excellent and the absence of heavy traffic allows them to be maintained easily and kept looking at their best.

PVC, vinyl asbestos, linoleum, cork and rubber are all suitable and as a very cheap solution thermoplastic tiles are adequate, particularly the new vinylised thermoplastic with improved colour ranges.

Vinyl asbestos tiles run through from the dining room to the kitchen with a change of colour in alternate squares.
House at Gattenside, Scotland
Architect: Peter Womersley
Photographer: William Toomey

18th century
Norfolk pammets
in a kitchen in
Hertfordshire.
Architects:
Derek Phillips
Associates

Heather brown
quarries in a
dining room in
Nottinghamshire.
Architects:
Diamond Redfern
and Partners
Photographer:
W. E. Middleton
& Son Ltd

Kitchens

The choice of flooring for the kitchen is the most difficult. It is the area of greatest wear and of maximum spillage.

John Prizeman in *Kitchens* *P.31 stresses the comfort aspect and considers 'resilience' of prime importance so that if china is dropped it may not break.

**Design Centre Publication* in this series

The 'Which' Report on Kitchen flooring (August '61) stresses the following aspects.

1 Easy to keep clean
2 Hard wearing
3 Not slippery
4 Stand up to water, oil, heat, abrasives, chipping, cracking, and denting
5 Look bright and cheerful
6 Not too cold
7 Quiet
8 Resilient

but then goes on to investigate linoleum and PVC types only—of these two types little advantage was found one over the other for most criteria, but linoleum was more affected by alkalis.

No perfect solution to the kitchen floor finish has yet been produced, and whatever is selected will be a compromise—durability at the expense of warmth, cheap first cost at the expense of long term appearance. It will depend upon your priorities. My own order of priorities would be first, durability, to ensure that the appearance and performance is not impaired by all the factors already listed which can be expected in a kitchen, which means that it must be hardwearing; and second appearance. I want a floor which gives an impression of quality related to all the other surfaces, colours and materials selected for the interior. Other factors which I consider of rather less importance are warmth, quietness and resilience, but a third important factor is ease of cleaning. I like a floor which doesn't immediately look filthy the moment someone with muddy shoes walks across it. My particular order of priorities has led to the solution of a slate floor with a riven surface, but other choices will be equally valid given a different order of priority. Other suitable floors include:

PVC sheet or tile, preferably with a resilient backing. Disadvantages—low resistance to heat

Ceramic tiles or quarry tiles: traditional kitchen flooring. Disadvantages—hard, cold

Ideally, a kitchen floor should be capable of having a bucket of water thrown on it and the floor mopped over – Prizeman suggests a central drain and in new buildings this may easily be accommodated; further, in order to protect walls, the floor finish should be coved up 3″ to 4″.

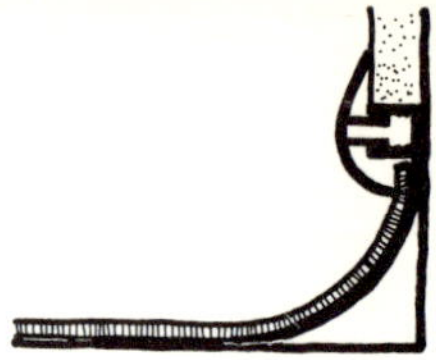

'Alucove' coving detail

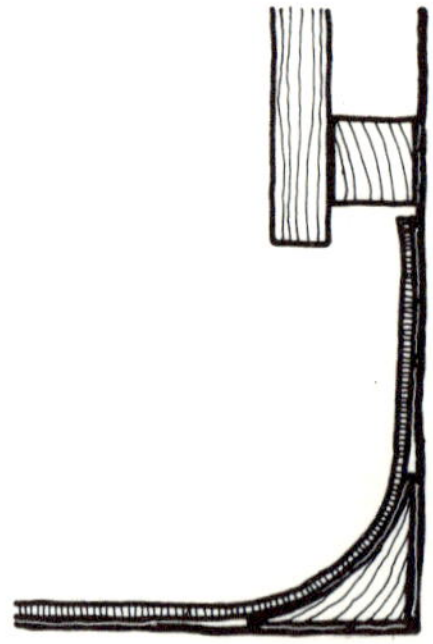

Cove formed with timber fillet

With flexible PVC floors this can be done by forming the finish up over a shaped wooden fillet.
For ceramic tiles special coved edges are available, but for stone, slate or quarry tiles, it is very expensive to have special coved edge sections made and a cheaper solution is to have cement coves formed, pigmented to match the floor colour.

Left:
Vinyl cork from kitchen to breakfast room at Nield House, Beaumond End, Buckinghamshire.
Architect: David Walters
Photographer: Jewell-Harrison

Right:
White mosaic in dining room and kitchen. Ferrum House (see page 31).
Architect: Jack Bonnington
Photographer: Henk Snoek

Grey vitreous tiles in the kitchen. Conversion of a house at Watford.
Architects: Derek Phillips Associates
Photographer: Jewell-Harrison

Playrooms

By playroom I don't mean a nursery solely set aside for the use of children—such a space is comparatively rare in homes now, and the nursery which doesn't after a few years change its use to accommodate teenage pleasures is rarer still. A playroom in this context then is an all-purpose family room, used for children at play, for teenage or adult parties as well as handicraft and listening to hi-fi. For other more purposeful pursuits see Peter Matthews' *Workrooms**.

The specification for the floor finish is, therefore, not an easy one. It should be durable—to take the heavy wear from children sliding and dancing, warm for children to sit on, and resilient enough not to be dangerous if a child falls and bumps his head. At the same time it should be easily cleaned as paint, ink, glue, wood shavings and water are likely at some time or other to find their way there, apart from the occasional childish 'accident'.

**Design Centre Publication.*

Resilient floors which are suitable include vinyl, cork and flexible PVC with resilient underlay, whilst linoleum, flexible PVC or vinyl asbestos tiles are less resilient alternatives.

Timber floors are a possibility, particularly hardwood block floors, but these will be more difficult to keep clean and of good appearance when water and other liquids are splashed about—it will be best to seal the floor rather than to attempt to maintain a good polished surface.

Rubber answers many of the needs but suffers from being insufficiently slippery for dancing, and too slippery when wet, so that it tends to be dangerous. Cork is likely to be difficult to maintain but is a good floor in other respects.

Cork tiles in a children's playroom at Shalford Hall, Essex.
Architect: Edward Samuel
Photographer: Stella Samuel

Bedrooms

Bedrooms, like living rooms, vary greatly in the extent and type of use—the night-time-only bedroom, the study bedroom or bed-sitting room. The bed-sitting room clearly has to be solved as a living room, the study bedroom, and this constitutes the majority of children's bedrooms in adequately heated houses, must work well as a bedroom but must also have a floor capable of withstanding paint, modern glues (model making), ink (homework) and dirt.

The essential quality in a bedroom is comfort, associated with warmth to bare feet. In our climate this will be provided by carpet. In the night-time-only bedroom when the amount of wear will be minimal, cheaper carpets are adequate. Dorothy Meade in *Bedrooms** P.54, makes the point that sisal or haircord are 'a bit prickly for bare feet', and clearly this will not give an impression of luxury. However, for study bedrooms, a good quality haircord will make a good compromise.

In bedrooms for small children the accent must be on ease of cleaning. As Claire Rayner points out in *For Children** P.13, a pile carpet isn't easy to clean 'after a child has been sick on it'. In nurseries the warmer and more resilient sheet or tile floor finishes such as cork or PVC with cork or felt backing are excellent, whilst lino, vinyl asbestos or thermoplastic are economic solutions.

One solution often illustrated in the glossy magazines is to have the floor boards sanded, polished, or sealed. Photographs of rooms in which this has been done, associated with brightly coloured carpet squares or rugs, can be very persuasive; but experience shows that softwood boarding is difficult to maintain, and whilst exposed boards may be satisfactory in areas of rooms which have little traffic, in areas of normal use they very soon begin to look scruffy.

Plain carpet in bedroom at Petersen House, Eynsford, Kent. Architect: Trevor Dannatt Photographer: Edgar Hyman

Bathrooms and WC's

This is an area where comfort is of utmost importance. The flooring should be warm and unslippery (see *Bathrooms* by Gontran Goulden* P.57).

**Design Centre Publication*

The choice will depend upon family composition; it will depend also on whether the bathroom contains a WC.

One solution is carpet. It answers the needs of comfort and gives an air of luxury, but if carpet is used it is important to choose the type of fibre which will not deteriorate, as even in the best kept single occupant bathroom, it will occasionally get splashed.

This is a good use for thick pile nylon carpet, and as a cheaper solution the non-woven types of nylon. Carpet is not the solution to the family bathroom with children kicking off their dirty shoes, and particularly if the bathroom contains a WC. There is a school of thought which says that WCs should be lower for medical reasons, and as pans are set lower it becomes more necessary for hygienic reasons to provide an unobtrusive urinal for men. However, for this reason in bathrooms containing WCs, overall carpet is not ideal, although large rugs might be.

A most luxurious solution is marble or ceramic tiles with underfloor heating; there are many exciting shapes and patterns and colours in clay tiles; and provided that these are warm in winter, the coolness in

Carpet run through from master bedroom to bathroom.
Nield House, Beaumond End, Buckinghamshire.
Architect: David Walters
Photographer: Jewell-Harrison

summer is an advantage. The only danger here is from slipping and this can be got over either by a non-slip finish, or an area of rug immediately adjacent to the bath. Unlike a kitchen, a bathroom floor has to be resistant chiefly to wet, so that any floor which doesn't suffer from water and isn't slippery can be used. PVC, preferably with resilient backing, vinyl asbestos tiles, cork or linoleum are all suitable but some will give a greater impression of warmth than others, and PVC, vinyl asbestos and linoleum are all slippery when wet.
Sheet materials can be formed up to provide coves at the edges to assist when someone leaves the bath overflowing, and a drain with levels laid to falls is an added refinement in the larger bathroom, or in houses where if water were to penetrate the ceilings of the rooms below, damage to valuable decorations would result.

Circulation Spaces

1 Corridor, halls: Circulation areas of a home, by definition, will get a lot of wear, so that a floor finish must be chosen for durability; it should also be easy to clean and, particularly on bedroom levels, should be quiet.
The area of most dirt and wear will be at entrances and there is a good case for the provision of a mat sinking at this position. This can be formed using a metal surround in jointless floors to protect the edges or if the edges of the finish are liable to cracking. If, however, ceramic or stone floors are used, a simple method is to form the recess by stopping the tiles or slabs, allowing sufficient depth for the particular type of mat to be provided. This can be the normal brush mat, for which at least a $1\frac{1}{2}''$ recess is required or, alternatively, ribbed rubber sheet which requires only about $\frac{3}{8}''$, or matting $\frac{3}{4}''$.

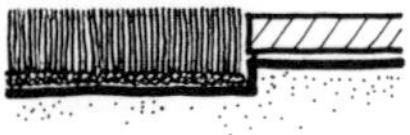

Mat sinking formed in quarries.

Standard sizes of brushmats are as follows, and sinkings should always be made approximately $\frac{1}{2}''$ oversize: $14'' \times 24''$, $16'' \times 27''$, $18'' \times 30''$, $20'' \times 33''$, $22'' \times 36''$, $24'' \times 39''$, $26'' \times 42''$, $28'' \times 45''$, $30'' \times 48''$. Non standard mats can be made up but these are likely to be difficult to get, and more expensive than equivalent standards.
Lower circulation areas are a good opportunity to use hard floors which satisfy the criteria for good looks and durability whilst being possible with solid floor construction. Excellent for this purpose are ceramics such as clay floor tiles, quarries, mosaics or bricks, concrete or stone. Upper circulation areas, associated with suspended floor construction, can use cork, rubber, or carpet, but if carpet is used it should be medium to good quality. A good solution is to run carpet throughout the whole of the first floor, covering bedrooms and corridors, with the possible exception of bathrooms. A good quality tufted carpet would be a solution, but clearly this will be expensive, and some compromise might be made with a similar looking but less durable and expensive carpet in the bedrooms where the amount of wear will be less.
2 Stairs: The majority of houses have timber softwood boarded staircases and as this is an unsuitable material for a wearing surface,

some form of floor finish is necessary. This finish should be hard wearing and should not show the dirt it will inevitably pick up. If noise is a consideration then carpet will be the best solution, and if carpet is used, it is as well to choose one of medium to good quality, and preferably one of the wool mixtures strengthened by nylon. Wilton, Axminster, or a good quality tufted will give good service, but cords tend to look threadbare after a while, although they are not worn out. An extra 18″ in length allows the carpet to be shifted occasionally to even out the wear. Good quality underlays should always be used in pads above each tread. Flexible PVC, with resilient backing, or sisal are adequate and inexpensive alternatives.

'Open riser' timber tread staircase, with terrazzo tile floor. House in Hampstead. Architect: Jellicoe and Colleridge Photographer: de Burgh Galway

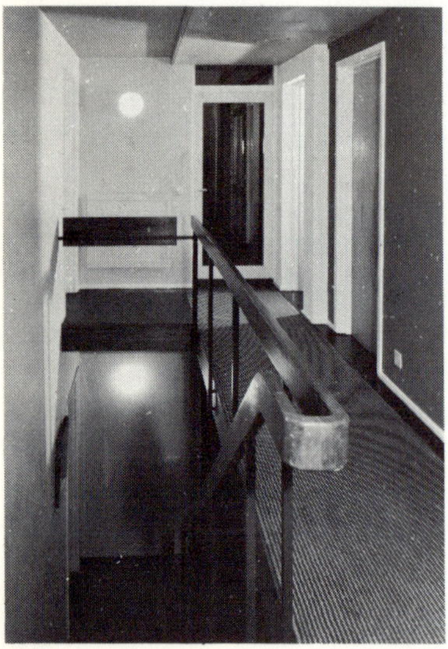

Left: Sisal carpet on bedroom landing. House in Hampstead. Architect: Trevor Dannatt Photographer: Colin Westwood

Right: Non woven nylon carpet inset into timber tread 'open riser' staircase. Flooring is sapele block parquet. Architect: Penelope Whiting Photographer: Richard Einzig

Secret fixing, stair carpet

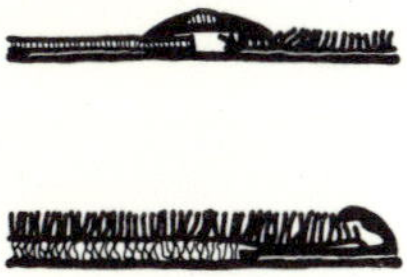

Cover strips at change of floor surface

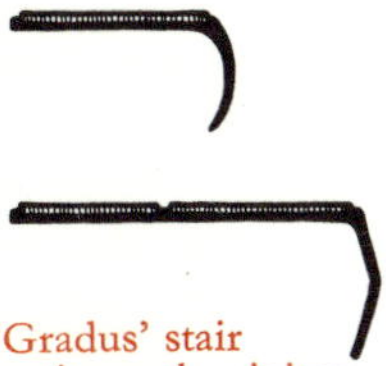

'Gradus' stair nosings, aluminium

'Marley' stair nosing, plastic

If possible, carpet should be taken the full width of the staircase to avoid the difficult task of cleaning the timber left exposed at either side. Special concealed methods of fixing are available.

Some houses have open staircases with hardwood treads. These can be left without a floor finish and will wear well, but if they are to be kept looking good they should be sealed and will require to be kept clean and free from abrasive dirt. Too much polish will make them slippery and dangerous but small quantities of non-slip polish may be used. As hardwood treads tend to be noisy, carpet may be used with them, but if wrapped around the tread looks clumsy—a neater method is to so profile the tread as to leave $\frac{1}{4}''$ recessed to accept a stuck non-woven nylon type of carpet which will give good wear and cut down noise.

In flats or maisonettes concrete stairs are sometimes found, and these can have a granolithic topping to give a hardwearing surface. Alternatively, thin sheet materials such as PVC, vinyl asbestos, linoleum, cork or rubber can be stuck to the risers and treads, and manufacturers provide a range of specially shaped nosings to cover the joint between the front of the top of the riser and the tread.

Concrete steps with granolithic or most thin sheet finishes tend to be noisy, but if noise is not a serious problem, special shaped clay floor tiles or quarries are available to give a really hard wearing finish.

12″ square quarries in corridor access to en suite spaces at house at Haddenham, Buckinghamshire. Architect: Peter Aldington Photographer: Richard Einzig

Exterior Spaces

1 Garages may not even have a floor finish. The concrete slab may be left with a well trowelled surface or at best a screed or granolithic topping may be added.

Very few people take much pride in a garage floor, and those that do, feel it sufficient to put down a box of wood shavings below the more 'leakable' parts of their car.

Concrete has low resistance to acids, and whilst having a high resistance to mineral oils, it will stain and look very dirty.

A much better alternative, although more expensive, will be quarry tiles, since they can be cleaned easily. It may be tempting to run the tarmac drive or asphalt surface into the garage, but this should be resisted as oils cause asphalt to deteriorate quickly.

It is inexpensive to install a circuit of underfloor electric heating in a garage, designed to be controlled by thermostat to operate only on the coldest nights to maintain the garage above freezing level. It will cost a negligible amount to run. This will operate successfully with quarry tile or granolithic concrete, and needs a little thought rather than a long pocket.

2 Patios and Terraces The first essential is that any floor finish should be frostproof, and this immediately limits the field of choice to:

Asphalt
Ceramics—fully vitrified types
Concrete granolithic tiles
Stone

There is available a non-woven polypropylene carpet designed to withstand snow or ice conditions and for use on patios and around swimming pools, but there is little experience of its use at present.

One of the exciting aspects of contemporary building is the close relationship which often exists between interior and exterior space.

Large floor to ceiling windows can be designed to slide or fold away to extend the living space outwards to the garden or terrace.

It is important, therefore, in selecting floor finishes which are related in this subtle way so that the one will enhance the other.

It is possible to use fully vitrified clay floor tiles or quarries both inside and outside so as to contribute to the feeling of spatial continuity.

Stone flagstones, particularly sandstones such as York Stone, make beautiful terraces and the slight irregularities of old used stone allow rock plants to grow between. Concrete flagstones, such as Noelite, provide a more economic solution, and these can be obtained in a variety of regular or irregular shapes in a limited number of colours. Colour should be used sparingly as it is difficult to compete with nature's colours, and to avoid bizarre effects. The natural colour of brick paving never jars and some of the finest effects can be achieved using bricks, or bricks combined with stone or concrete flags.

For floor finishes around swimming pools, fully vitrified clay floor tiles are produced with non-slip finishes, with all the necessary fittings to enable necessary gulleys and bull nosed edges to be formed.

12″ square quarries carried through from interior to exterior patio. House at Haddenham, Buckinghamshire (see page 41). Architect: Peter Aldington Photographer: Richard Einzig

Hardwood strip floor carried through from interior of living room to covered balcony. House at Beaulieu. Architect: Sir Basil Spence Photographer: de Burgh Galway

Types of Flooring

Asphalt

MASTIC ASPHALT

A material based on natural rock as fillers such as limestone. It is laid hot to form a waterproof floor, either jointless or moulded into tiles. Special acid resistant types are available. $\frac{5}{8}''$ thick jointless floors are suitable for domestic use. Not much used now due to surface softening with oils and grease. Colours: dark red, brown or black. Matt or polished finishes.

	Subfloor	Jointless/Size	Thickness	Manufacturers
BS 1410 (Jointless)	Jointless: Any sound and rigid subfloor	$9'' \times 4\frac{1}{2}''$ $8'' \times 8''$	$\frac{5}{8}'' - 1\frac{1}{2}''$	J. Cooke & Son (Huddersfield) Ltd
BS 1076/1451	Tiles: Concrete sand screed	$10'' \times 10''$ $8'' \times 4''$	$\frac{5}{8}'' - 2\frac{1}{2}''$	Limmer & Trinidad Co Ltd
BS 1324 (Tiles)		$5\frac{1}{2}'' \times 5\frac{1}{2}''$		

PITCH MASTIC

Made from coal tar pitch with fillers such as limestone. It has characteristics similar to Mastic Asphalt and is laid hot. It has a higher resistance to oils and fats, but lower resistance to indentation than Mastic Asphalt. $\frac{5}{8}''$ thickness suitable for domestic use. Colours: dark black, red or brown. Matt or polished finishes.

	Subfloor	Jointless/Size	Thickness	Manufacturers
BS 1450	as for Mastic Asphalt	Jointless	$\frac{5}{8}'' - 1''$	Limmer & Trinidad Co Ltd Permanite Ltd
BS 3672 (Coloured)				

Carpets

The main fibres used in the pile are wool, nylon, acrylic and rayon, and these fibres are used on their own or mixed in various proportions. As it is difficult for the general public to know what trade names for synthetic fibres mean, the following list may be of help.

Nylon — Bri-nylon, Du pont 501, Enkalon, Allyn 707

Acrylic — Acrilan, Courtelle, Orlon Type 36

Rayons or Viscose — Fibro, Evlan, Evlan M

A table in which the properties of different fibres are illustrated, graded from A (best) down to D.

	Wool	Nylon	Acrylic	Rayon
Soil resistance	A	D	C	B
Strength	C	A	B	D
Resilience	A	C	B	D
Cheapness	B	D	C	A

Each fibre has advantages and disadvantages, which is the reason why mixtures have been developed to give a compromise and to try to maximise on the advantages.

WOOL Warm, resilient, wears well, does not soil easily. Does not ignite easily and is self extinguishing.

NYLON Tough, soils easily but is easily cleaned. Melts on exposure to flame.

ACRYLIC Warm and resilient, wears well, soils easily like nylon, but is easily cleaned. Flammable.

RAYON Cheap. Mainly used for bulk in mixtures for economy.

Mixtures
The following fibres are often used:

100% wool
100% nylon
100% acrylic
100% rayon
80% wool 20% nylon
80% wool 20% rayon
50% wool 50% rayon
42½% wool 42½% rayon 15% nylon
40% wool 40% rayon 20% nylon
40% acrylic 40% rayon 20% nylon
80% rayon 20% nylon

New mixtures are constantly being developed, as are new or improved fibres. It is always best to consult a carpet specialist, but when making a decision don't be misled by the high powered advertising. Bear in mind the characteristics of each material, check the fibre content and mixtures, and see if it will give the sort of service required. Probably the most successful mixture at present is 80% wool 20% nylon in medium grades. This has the overwhelming long-term advantages of wool, with added strength from the nylon content. The least useful will be 100% rayon. New materials for carpets are constantly being developed: for example 100% polyester fibre carpets are now available, and experimental work is being carried out with paper carpets and stainless steel fibre. A further interesting development is that of reinforcing any type of carpet with resin to provide increased durability.

AXMINSTER

Axminster is a woven carpet, in which the pile is placed in position at the same time as the backing is woven. It varies from Wilton in the way in which the pile is retained, the method allowing an unlimited number of colours. Axminsters in general have a less dense pile than Wilton but greater variety of designs. The pile is always cut. Few plain woven carpets are Axminster. A large number of colours and patterns are available and new patterns are constantly being developed.

	Subfloor	Jointless/Size	Thickness	Manufacturers from Design Index
Broadloom	Any dry even subfloor with recommended underlay	9′ 12′ 15′ wide Length up to 25–30 yds	¼″ – ¾″	W. & R. R. Adam Ltd Broadloom Carpets Ltd The Carpet Manufacturing Co Ltd
Narrow	Jointed carpet/felt. Broadloom/felt or foam types	27″ 36″ wide Length up to 50 yds		John Crossley & Sons Ltd Andrew Gaskell Ltd Gilt Edge Carpets Ltd John Lyle & Co Ltd S. J. Stockwell & Co (Carpets) Ltd A. F. Stoddard & Co Ltd James Templeton & Co Ltd Tomkinsons Ltd Ulster Carpet Mills Ltd

Spool Axminster weave

'Canberra'. Axminster carpet by Broadloom Carpets Ltd. Designer: Michael Coo

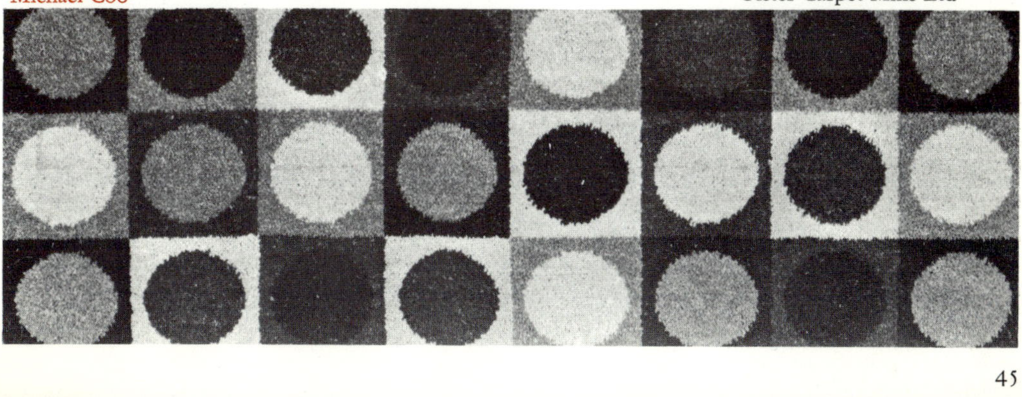

WILTON

The principal difference between Wilton and Axminster carpet is that all colours used in the pile are woven into the carpet as continuous strands. Due to this the colours tend to be more limited. Plain woven carpets are generally Wiltons. Special designs can be obtained economically for lengths of carpets over 100 linear yards.

	Subfloor	Jointless/Size	Thickness	Manufacturers from Design Index
Broadloom	as for Axminster	9′ 12′ 15′ wide	$\frac{1}{4}'' - \frac{3}{4}''$	Blackwood Morton & Sons Ltd
				Bond Worth Ltd
Narrow		27″ 36″ wide Length up to 50 yds		Brintons Ltd
				Buckingham Carpet Co Ltd
				The Carpet Manufacturing Co Ltd
				John Crossley & Sons Ltd
				Firth Carpets Ltd
				Gilt Edge Carpets Ltd
				Grays Carpets Ltd
				Heckmondwike Carpets Ltd
				Hugh Mackay & Co Ltd
				The Old Bleach Linen Co Ltd
				I. &. C. Steele & Co Ltd
				S. J. Stockwell & Co (Carpets) Ltd
				A. F. Stoddard & Co Ltd
				James Templeton & Co Ltd
				Thomson, Shepherd Carpets Ltd
				Tribute Carpets
				The Wilton Royal Carpet Factory Ltd

Wilton weave

TUFTED

The chief difference between Axminster/Wiltons and 'tufted' is that the pile is stitched into an already prepared jute backing and not woven with it. The pile is secured by a latex backing. The pile can be either looped or cut, and until recently patterns have been limited by the process to self colours and mottled textures, or 'cut-in' patterns. However, two or three colour combinations have now been introduced by a more sophisticated manufacturing process. At first the fibres generally used were rayon pile for cheapness, but as these wore badly it is usual to reinforce this with wool or nylon mixtures, 100% nylon and 100% acrylic is also used.

	Subfloor	Jointless/Size	Thickness	Manufacturers
Broadloom	as for Axminster	27″ – 15″ in variety of widths to suit most room dimensions. Lengths up to 50 yds	$\frac{1}{4}'' - \frac{1}{2}''$	Barwick Carpets Ltd
				Georgian Carpets Ltd
				Holmbury Carpets
				Kosset Carpets Ltd
				Lancaster Carpets Ltd
				Rivington Carpets Ltd
				Shaw Carpet Co Ltd
				Velmar Textiles Ltd

CORDS

These carpets are manufactured in a similar manner to Wilton, but the uncut pile is looped tightly over the backing giving cords their characteristic lines, and directional appearance. Other methods of production are also used. Animal fibres such as goats hair, are used for 'hair cords', and these are particularly hard wearing. Cheaper fibres such as rayon are also used. Wide range of colours available.

	Subfloor	Jointless/Size	Thickness	Manufacturers
Broadloom	as for Axminster	27″ – 15′ Lengths up to 50 yds	$\frac{1}{4}''$	John Crossley & Sons Ltd
Narrow				Firth Carpets Ltd
				A. F. Stoddard & Co Ltd
				Trafford Carpets Ltd
				Tretford Carpets Ltd
				Wiltshire Carpets Ltd

1

2

3

1 'Muggins Clover' Wilton carpet by S. J. Stockwell (Carpets) Ltd Designer: Olive Sullivan

2 100% acrylic tufted carpet by Barwick Carpets Ltd

3 80% wool and 20% nylon cut pile carpet designed for the Queen Elizabeth 2 by Kosset Carpets Ltd Designer: Jon Bannenberg

To give detailed information of all types of carpets and mats would be impossible in a general book on floor finishes. Here are other types, some of which are available as either carpets or rugs.

		Manufacturers
Needleloom felt	A comparatively new type of floor covering in which nylon or other fibres are needled into a jute hessian back, designed to be stuck to an even subfloor. Polypropylene versions can be used externally. Hardwearing	Armstrong Cork Co Ltd Dermide Ltd Dunlop Semtex Ltd Heckmondwike Carpets Ltd E. Illingworth & Co (Bradford) Ltd Marling Industrial Felts Ltd Barry Staines (Sales) Ltd R. & B. Textiles Ltd
Nylon with PVC backing	100% nylon cut pile with PVC backing stabilised with glass fibre. Hardwearing	Debron Carpets Ltd
Plush carpets	100% nylon pile velvet fabric on cotton backing with integral foam underlay	Shildon Carpets Ltd
Carpet tiles	Interchangeable animal hair pile carpet tiles, approx 20″ × 20″. Wide variety of colours. Tiles amalgamate together when down but can be replaced individually	Tretford Carpets Ltd Van Heugten Bros Ltd
Cotton	Dense soft smooth pile. Hardwearing but soil easily	(Generally unbranded)
Indian	Hand made, with the pile looped and knotted into the backing. Good quality gives reasonable wear, avoid cheap ones—tendency to flatten and to fluff	(Generally unbranded)
Numdah rugs	Indian rugs with embroidery on a felt base. Poor durability. Cheap	(Generally unbranded)
Rush mats	Avoid very dry conditions	The Waveney Apple Growers Ltd

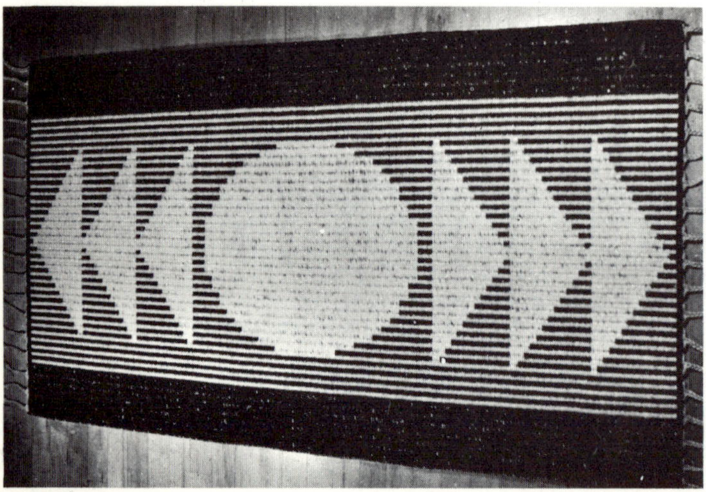

Left:
Reversible flat woven rug in black and white, 3′ × 5′, by Peter Collingwood

Right:
100% wool Rya rug by Firth Carpets Ltd

Axminster Rya rugs	Usually patterned. Long pile flattens easily.	Broadloom Carpets Ltd Bond Worth Ltd Firth Carpets Ltd Grays Carpets Ltd John Lyle & Co Ltd Quayle Carpets Ltd A. F. Stoddard & Co Ltd Tomkinsons Ltd
Handwoven rugs	Various types, most are comparatively expensive	Michael Abakhan Blackfriars Settlement Bronte Tapestries Geraldine Carmichael Peter Collingwood Highland Home Industries Ltd Inverhouse Handlooms Robert Mabon Barbara Mullins Gwen Mullins
Woven PVC rugs	Similar wearing properties to flexible PVC sheet or tiles. Easy to keep clean, proof against staining. Cheap	W. Armes & Son Ltd The Belfast Ropework Co Ltd* British Ropes Ltd William Goodacre & Sons Ltd Green Brothers (Geebro) Ltd Jute Industries Ltd *woven orientated polypropylene

Not mentioned above but generally available are differing types of plain rugs, oriental (Persian, Afghan, Chinese) rugs, scandinavian long pile and flat woven rugs and a wide variety of exotic kinds from different countries.

Ceramics

CLAY FLOOR TILES

Made from refined natural clays, pressed into tiles and fired at high temperature. Fully vitrified tiles have additives and are fired to higher temperatures. The fully vitrified types are frost proof and can be used externally. Clay tiles are made with a wide variety of colours, surface finishes (glazed or natural), patterns and textures, and are impervious to all normal household fluids. They are particularly useful in kitchens Some are non-slip.

	Subfloor	Jointless/Size		Thickness	Manufacturers
BS 1286/B	Concrete with	3″ × 3″		$\frac{3}{16}$″ $\frac{1}{4}$″ $\frac{3}{8}$″	Carter Tiles Ltd
	or without	4″ × 4″		$\frac{1}{2}$″ $\frac{5}{8}$″ $\frac{3}{4}$″	H. & R. Johnson Ltd
	screed	$4\frac{1}{4}$″ × $4\frac{1}{4}$″		$\frac{7}{8}$″	Langley London Ltd
		6″ × 6″			Pilkington's Tiles Ltd
		also			Daniel Platt & Sons Ltd
					Richards Tiles Ltd
		$4\frac{1}{4}$″ octagonal			George Woolliscroft & Son Ltd
		6″ × 1″	6″ × 3″		*Imported Tiles*
		6″ × 4″	6″ × $4\frac{1}{2}$″		E. Eaton Wholesale Ltd
		8″ × 4″	8″ × 5″		Verity-Orient Tiles Ltd
		7″ × 7″	8″ × 8″		Dennis M. Williams Ltd
		$9\frac{1}{2}$″ × $4\frac{1}{2}$″	9″ × 9″		
		10″ × 10″	12″ × 10″		
		12″ × 12″	16″ × 16″		

QUARRIES

The word 'quarry' comes from the French word carré meaning square. Quarry tiles are extruded from unrefined clays, and hard burnt. They can be used externally, but only those of first quality. Second quality may delaminate. The larger sized quarries are more difficult to obtain and are more expensive. Quarries have been traditionally used in kitchens, but they have many other uses. Quarries are very hard wearing, and the colours have a random quality. Colours: Natural browns, buffs and reds, also dark blues.

	Subfloor	Jointless/Size	Thickness	Manufacturers
BS 1286/A	Concrete and	9″ × 9″	$1\frac{1}{4}$″ $1\frac{1}{2}$″	John Caddick & Son Ltd
	screed	6″ × 6″	$\frac{5}{8}$″ $\frac{7}{8}$″	Langley London Ltd
		4″ × 4″	$\frac{1}{2}$″ $\frac{3}{4}$″	Daniel Platt & Sons Ltd
				Dennis Ruabon Ltd
		also		Wheatley & Co Ltd
		8″ × 8″	$1\frac{1}{8}$″	
		12″ × 12″	$1\frac{1}{2}$″ 2″	

MOSAIC

(various types)

Manufacture of mosaics is as for normal clay floor tiles, and only the size is different. In addition, however, glass mosaic is available, also ceramic mosaic on a rubber backing and marble mosaic. They have similar characteristics to clay floor tiles and give a rich appearance. Floors should be coved at the edges where possible. Mosaic floors can be obtained in a wide variety of colours and sizes. They can be either glazed or unglazed, semi or fully vitrified. Random patterns are available.

Subfloor	Jointless/Size	Thickness	Manufacturers
Concrete and	$\frac{3}{8}$″ × $\frac{3}{4}$″	$\frac{3}{16}$″ – $\frac{1}{4}$″	Carter Tiles Ltd
screed (timber	2″ × 2″		Drama Decor Ltd
boards can be			Jason Mosaics Ltd
made suitable	many intermediate		H. & R. Johnson Ltd
with ply or	sizes		Langley London Ltd
hardboard			Pilkington's Tiles Ltd
underlay for	1″ hexagonal		Richards Tiles Ltd
rubber backed			Runnymede Rubber Co Ltd
mosaics)	also 2″ × 1″		
	(generally		
	paper mounted		
	for easy		
	installation in		
	2′ × 1′ or 2′ × 2′		
	square sheets)		

BRICK

Most types of hard brick are suitable for flooring. They may be used both internally and externally and are particularly suitable for use in strips close to windows opening out on to terraces where the same material can run through. Brick paving has the advantage that it can easily be formed to take up curves or irregular shapes. The appearance will depend on type of brick used, but these will usually be of the darker type – blues, purples, buffs and browns.

Subfloor	Jointless/Size	Thickness	Manufacturers
Concrete internally. Externally laid in sand bed or weak cement on levelled or rammed earth or hardcore	$8\frac{5}{8}'' \times 4\frac{1}{4}''$	$2''$ $2\frac{5}{8}''$ $2\frac{7}{8}''$	Redland Bricks Ltd

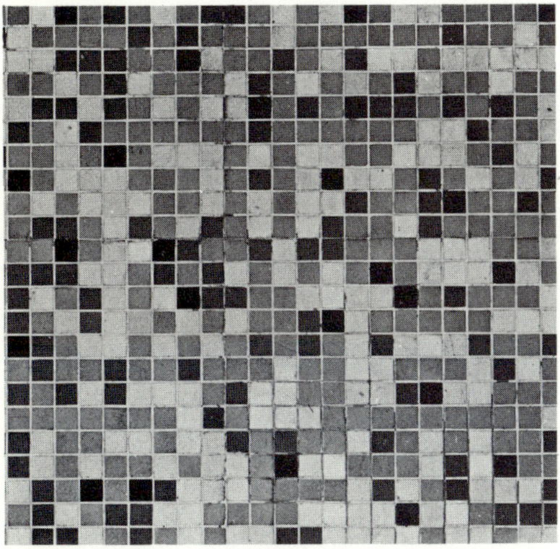

Mosaic by H. & R. Johnson Ltd

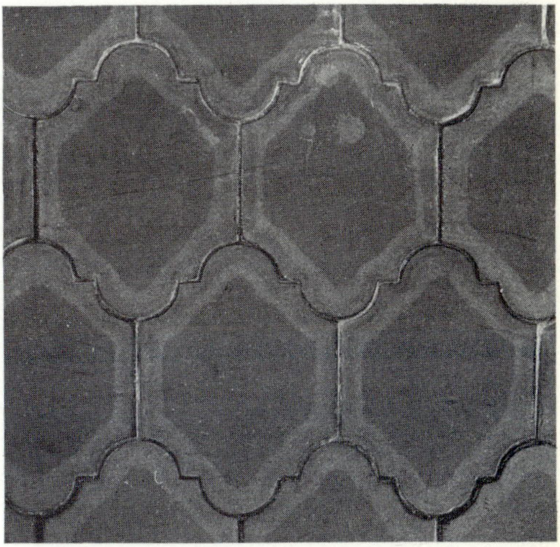

French 'Provencal' clay tiles imported by E. Eaton Wholesale Ltd

Heather-brown quarry tiles by Langley London Ltd

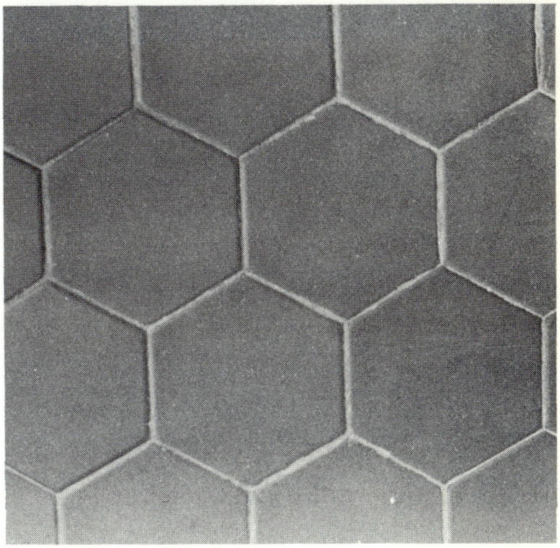

Hexagonal clay tile paviors by Daniel Platt & Sons Ltd

Concrete

GRANOLITHIC

Cement mixed with graded granite chippings to form an in-situ topping – this should be laid integrally in one or two coats with the concrete base, by a specialist. Joints should be provided over wide areas to avoid cracking (approx 12′×12′ without joints, or 12′×24′ of monolithic toppings). A monolithic 'grano' finish is obtained by laying the topping while the subfloor is still in a plastic state (within 3 hours of laying). An excellent hardwearing and dustproof finish is obtained. Granolithic finishes can be laid to a limited range of colours.

Subfloor	Jointless/Size	Thickness	Manufacturers
Matured or new concrete. (Matured concrete can be used with special precautions)	Bays not exceeding 240′ super	$\frac{1}{4}''$ – 1″ monolithic 1″ – 3″ as separate topping	Johnson Floor Co Ltd

Concrete pavior by Noelite Ltd

CONCRETE TILES

(and precast flags)

Concrete tiles are formed with coloured cement and hardwearing aggregate surface on a normal concrete backing, to give a very hardwearing finish. Special carborundum finishes are available. Precast flags have a coarser texture and are used externally. Colours: Natural concrete and pale shades of grey and reds.

	Subfloor	Jointless/Size	Thickness	Manufacturers
Tiles BS 1197	Concrete	Tiles 4″×4″ – 20″×20″ in a range of sizes standard sizes 9″×9″ 12″×12″ 16″×16″ 20″×20″	$\frac{5}{8}''$ – $1\frac{1}{2}''$ $\frac{7}{8}''$ $1\frac{1}{8}''$ $1\frac{1}{4}''$ $1\frac{1}{4}''$	E. J. Elgood Ltd Haskel, Robertson & Co Ltd Langley London Ltd The Marley Tile Co Ltd Noelite Ltd Prodorite Ltd
Flags BS 368	Sand or weak cement bed on hardcore or well rammed earth	Precast flags 2′×1′6″ 2′×2′ 2′×2′6″ 2′×3′ 27″×18″ wide variety of intermediate sizes	 $1\frac{1}{2}''$ 2″ – $2\frac{1}{2}''$	

TERRAZZO/JOINTLESS/TILES

A special form of concrete using cement with decorative aggregates. The aggregate gives the pattern and colour, and contains varying sizes of marble chippings to give coarse or fine patterns. Jointless terrazzo is rarely used domestically, but Terrazzo tiles are a useful addition to the range of hard floor finishes available. A very wide range of patterns and colours are available.

	Subfloor	Jointless/Size	Thickness	Manufacturers
Tiles BS 4131	Concrete with screed	Jointless up to 12 sq ft maximum panel	$\frac{5}{8}$″ – $1\frac{1}{2}$″	Diespeker & Co Ltd J. Joslin & Co Ltd Jaconello Ltd Marble Mosaic Co Ltd
	Concrete with screed	6″ × 6″	$\frac{5}{8}$″	A. Quiligotti & Co Ltd
		8″ × 8″	$\frac{3}{4}$″	Roman Mosaic Ltd
		9″ × 9″	$\frac{3}{4}$″	
		12″ × 12″	$1\frac{1}{8}$″ & 1″	
		16″ × 16″	$1\frac{1}{4}$″	
		larger tiles up to 36″ × 36″ available	2″	

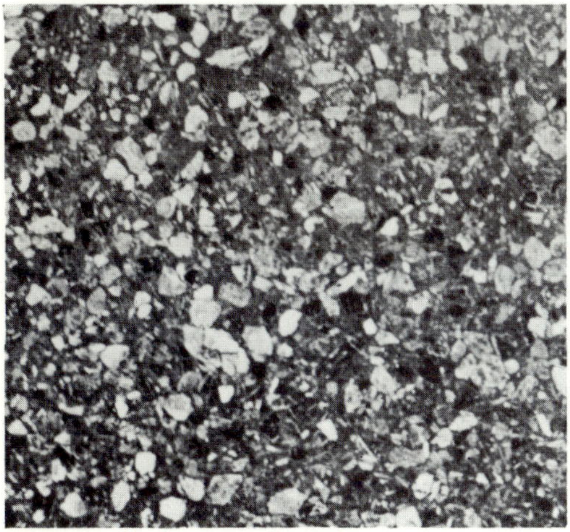
Terrazzo tiles by Jaconello Ltd

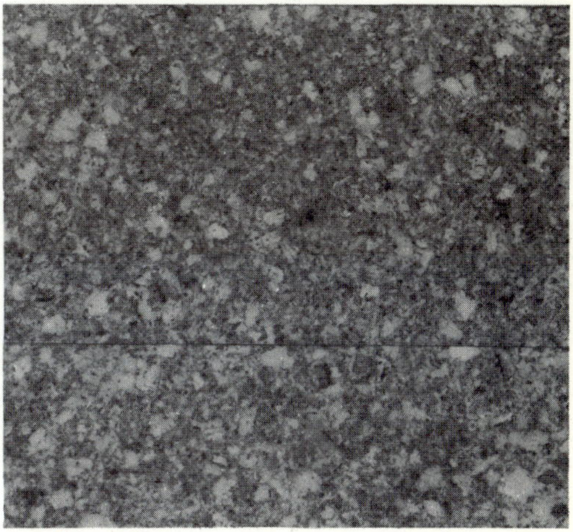
Vinyl cork by Wicanders (Great Britain) Ltd.

Cork

Natural cork granules compressed and baked to form tiles, bonded together either by the natural resins or by additional resins to permit lower temperature moulding. Warm resilient flooring, very quiet tread. Good wearing properties, particularly when sealed with polyurethane. Cork coves and fittings available. Vinyl cork has a thin vinyl skin surface which increases its durability, and should not be sealed. 9″ × 9″ available. Colours are natural cork, light, medium or dark, also dyed in a number of colours.

	Subfloor	Jointless/Size	Thickness	Manufacturers
Cork tiles	Any dry even subfloor	6″ × 12″	$\frac{1}{8}$″ $\frac{3}{16}$″	Armstrong Cork Co Ltd
		9″ × 9″	$\frac{5}{16}$″	British Technical Cork Products Ltd
Vinyl cork		12″ × 12″		
	Concrete/screed	12″ × 24″		Mundet Cork and Plastics Ltd Robinson Bros Cork Growers Ltd
	Timber/hardboard or ply underlay			Wicanders (Great Britain) Ltd
	Tiles pinned and stuck			

Linoleum

JUTE/HESSIAN BACKING SHEET OR TILES; INLAID

A flooring prepared from linseed oil, resins, cork and wood flour, calendered on to a jute/hessian backing. A hard wearing reasonably resilient and warm floor. Widely used domestically, but now giving way to cheaper grades of carpet. It can be coved up at the edges for use in bathrooms or kitchens. Special indentation resistant grades available to combat stiletto heels. Inlaid linoleum consists of patterns of different colours, wood grains and imitation wood strip or parquet. Very wide range of plain or mottled colours available.

	Subfloor	Jointless/Size	Thickness	Manufacturers
BS 810	Any dry even subfloor	Sheet 72″	1.6mm	Armstrong Cork Co. Ltd.
		Strips 4″ × 36″	1.8mm	Barry Staines (Sales) Ltd
		Tiles 9″ × 9″	2.0mm	The Dundee Linoleum Co Ltd
	Underlay required over boards/boards to be ventilated	12″ × 12″	2.5mm	Limmer and Trinidad Co Ltd
		18″ × 18″	3.2mm	Nairn Floors
		24″ × 24″	4.0mm	Tayside Floorcoverings Ltd
		Lengths 25-30 yards	4.5mm	
			6.0mm	
			6.7mm	

Cork backed stiletto resistant linoleum by The Dundee Linoleum Co Ltd

SATURATED PAPER BACKED; FELT-BACKED; PRINTED FELT-BASE

Linoleums are also available known as 'felt-backed', usually in tile form. The main difference is in the backing, this is calendered on to a bitumen saturated paper felt instead of hessian. The cheapest type of all which is not a true linoleum is 'printed felt base' a material with a printed paint finish on a bituminous felt backing. A wide range of colours in plain or marbled types provided.

	Subfloor	Jointless/Size	Thickness	Manufacturers
BS 1863	similar to jute/hessian backing	Sheet 6′ printed felt base	1.0mm	Armstrong Cork Co Ltd
				Barry Staines (Sales) Ltd
		Tiles 9″ × 9″	2.5mm	Limmer & Trinidad Co Ltd
		12″ × 12″ felt backed	3.2mm	Nairn Floors

CORK CARPET

Cork carpet is a similar material to linoleum (rather than to cork). It is manufactured by the same process, but uses cork granules bonded together with linseed oils and resins. The surface is rougher then linoleum and tends to get dirty and is more difficult to clean. Not as hard wearing as linoleum. Should be sealed initially and given regular maintenance. Plain colours only.

	Subfloor	Jointless/Size	Thickness	Manufacturers
BS 810	similar to jute/hessian backing	Sheet only 6′ wide	3.2mm	Barry Staines (Sales) Ltd
			4.5mm	Nairn Floors
		Lengths 50′	6.0mm	

Magnesium Oxychloride

Often known simply as 'composition flooring', magnesite has been well known for many years in low cost housing. It has no moisture resistance, but can be protected by linseed oil and subsequent waxing. It is hard and glossy, but relatively warm. All pipes in contact must be protected by bitumen against corrosion. The only jointless floor finish suitable for timber boarded floors. Available in many colours including white.

	Subfloor	Jointless/Size	Thickness	Manufacturers
BS776	Any rigid and dry subfloor suitable	Jointless	$\frac{5}{8}''$	The British Doloment Co Ltd The Liotex Asbestos Flooring Co Ltd British Magnesite Flooring Co Ltd Marbello & Durus Ltd Oswald Toffolo & Co Ltd A. Quiligotti & Co Ltd

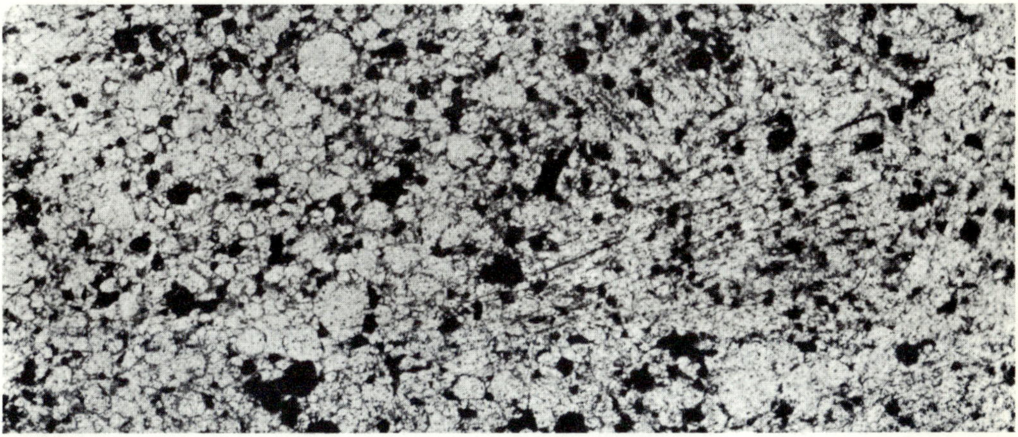

Magnesium oxychloride floor by the Liotex Asbestos Flooring Co Ltd

---

Plastics

THERMOPLASTIC TILES; VINYLISED THERMOPLASTIC TILES

Thermoplastic tiles (sometimes called asphalt tiles) are formed from asbestos fibres, mineral fillers and pigments, together with a thermoplastic binder. The darker binders and fillers are cheapest, giving the cheaper range of tiles. They form one of the largest used floor finishes (introduced 1946), but earlier types are being replaced by vinylised thermoplastic which with its vinyl content gives a degree of flexibility and is more resistant to abrasion and oils. Thermoplastic tiles have a limited range of dark flecked colours with greatly improved ranges in the vinylised types.

	Subfloor	Jointless/Size	Thickness	Manufacturers
BS2592	Concrete with screed DPC not absolutely necessary, but advisable if ground has tendency to be damp. Strong rigid timber floor with $\frac{1}{2}''$ underlay. DPC required in new construction by 1965 Building Regulations	$9'' \times 9''$ tiles	$\frac{1}{8}''$ $\frac{3}{16}''$ $\frac{1}{10}''$	Armstrong Cork Co Ltd Dunlop Semtex Ltd The Marley Tile Co Ltd

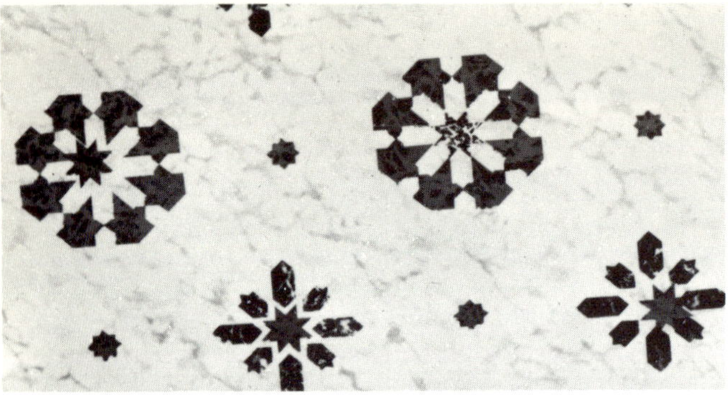
1

2

3

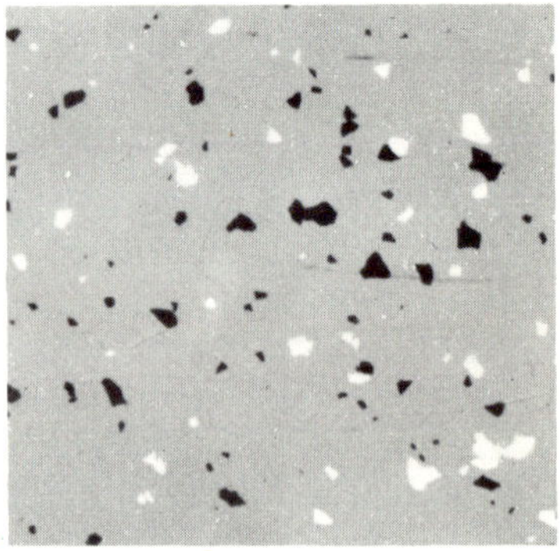
4

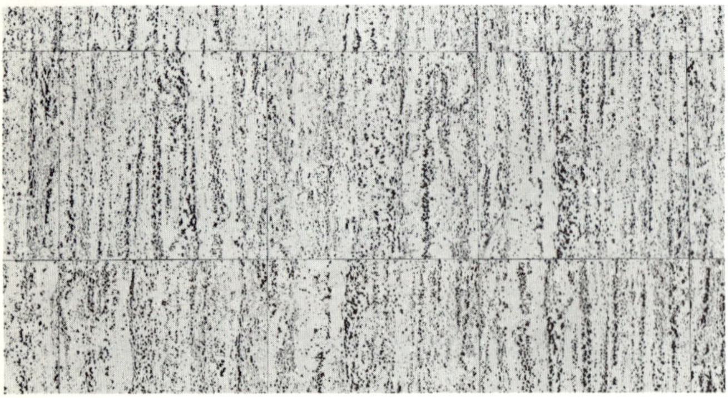
5

1 'Corinth' reverse printed PVC sheet flooring by Dunlop Semtex Ltd

2 'Vinyl Corlon' inlaid PVC sheet by Armstrong Cork Co Ltd

3 Thermoplastic tiles by Dunlop Semtex Ltd

4 Polyester resin floor by E. J. Elgood Ltd

5 Vinyl asbestos 'embossed' tiles by The Marley Tile Co Ltd

VINYL ASBESTOS TILES

A tile formed using PVC with fillers of asbestos fibres and limestone. Tiles are reasonably flexible, resist indentation, moisture, chemicals and oils. Cigarette burns can be removed with wire wool, tendency to mark with rubber heels. Hardwearing, and improvement on thermoplastic tiles in the low price range flooring. Wide range of clear colours, with flecked patterns.

BS3260	Subfloor	Jointless/Size	Thickness	Manufacturers
	Concrete with screeed even and dry	9″ × 9″	1.6mm	Armstrong Cork Co Ltd
		12″ × 12″	2.0mm	Dunlop Semtex Ltd
			2.5mm	Florco Ltd
	Timber floors with hardwood/plywood/chipboard underlay		3.2mm	The Marley Tile Co Ltd
				Nairn Floors

FLEXIBLE PVC SHEET/TILE TEXTURED OR EMBOSSED

Recent years have seen enormous advances in the field of PVC flooring. It is a most versatile and useful material. Plain surfaced sheet is available, or alternatively plain or textured or embossed tiles giving the appearance of tavertine marble, old ceramic tiles, mosaic or timber floors. A number of companies produce a high grade material consisting of vinyl chip set into a clear vinyl, giving the impression of mosaics or pebble floors. It is in the field of flexible PVC flooring that some of the promising developments are likely to take place. The material is easy to clean and hardwearing. Joints in sheet floors may be heat sealed to give a jointless floor. Colours: unlimited variety.

	Subfloor	Jointless/Size	Thickness	Manufacturers
BS3261	Any dry and very even subfloor. Timber must have ply or hardboard underlay	Sheet 36″ – 72″ wide Up to 30 yards long	1.1mm 1.6mm 2.0mm 2.5mm 3.2mm	Amtico Flooring Ltd Armstrong Cork Co Ltd Barry Staines (Sales) Ltd The Dundee Linoleum Co Ltd The Dunlop Co Ltd
		Tiles 9″×9″ – 10½″×10½″ 12″×12″ 18″×18″ 36″×36″		Dunlop Semtex Ltd Gerland Ltd James Halstead Ltd The Marley Tile Co Ltd Nairn Floors

PVC WITH HESSIAN BACKING; CORK BACKING; FELT BACKING; CELLULAR BACKING

Various materials have been bonded to the back of PVC sheet to give added resilience, dimensional stability and strength.

Hessian Greater strength for heavy traffic areas sold as sheet or tile.

Cork/felt Greater resilience and warmth, also acoustic insulation. Generally stuck to subfloor.

Cellular A cellular foam PVC structure laminated to flexible PVC sheet. Similar characteristics to cork or felt.

Variety of colours, patterns and textures, including wood grain.

	Subfloor	Jointless/Size	Thickness	Manufacturers
BS3261	Any dry and even subfloor	Hessian/sheet sheet 72″	2mm 3mm	The Adamite Co Ltd Armstrong Cork Co Ltd Dermide Ltd
		Cork sheet 57″ 78″ Cork/tiles 9″ sq 12″ 17¾″	2.7mm – 4.5mm	Dunlop Semtex Ltd James Halstead Ltd Hardura Ltd The Marley Tile Co Ltd Nairn Floors
		Felt/sheet 27″ 54″ 72″ 78″ wide	2.9mm 3.0mm 3.2mm 4.2mm	Sommer Developments Ltd Tayside Floorcoverings Ltd
		Cellular/sheet 48″ 60″ 72″	2.0mm 2.5mm 3.9mm	

PVC REVERSE PRINTED TYPES

A comparatively new process whereby any pattern can be reproduced photographically and printed in reverse on to a thin film of clear PVC which is then laminated to a PVC backing. The process is cheap and allows tremendous versatility of patterns. 25 million yards sold annually. The flooring is usually thin and will give a cold, hard floor. The wearing surface is of course very thin, generally not more than .1 or .2mm. One manufacturer has produced a nursery floor with patterns of games.

Subfloor	Jointless/Size	Thickness	Manufacturers
Any dry even and rigid subfloors	Sheet 48″	1mm – 1.5mm	Dunlop Semtex Ltd James Halstead Ltd (Playflor for nurseries) The Marley Tile Co. Ltd Nairn Floors

PVA EMULSION

A multi layer jointless floor, applied as a series of thin layers, based on polyvinyl acetate emulsion. It contains no cement and dries by evaporation. It is widely advertised as a do-it-yourself flooring, but it is not entirely suitable for interior domestic use, as it has low resistance to water. Colours: Plain and marbled.

Subfloor	Jointless/Size	Thickness	Manufacturers
Concrete, wood block, clay or stone – must be rigid and dry (Some manufacturers recommend its use on timber)	Jointless	$\frac{1}{32}$″ – $\frac{1}{8}$″	Laymat Flooring Co Surfex Flooring Co

---

POLYURETHANE RESIN; POLYESTER; EPOXY

Polyurethane, polyester and epoxy resins are all used to a limited extent domestically – mainly kitchens and wet areas. Polyurethanes are the least rigid and most adaptable to these circumstances. All are applied by specialists, and are applied by different methods, ie paint coats, trowel or roller coats, according to individual manufacturer. Ingredients (decorative) – coloured vinyl flakes, chips etc, giving plain, mottled and textured finishes. Polyurethanes with coloured flakes give an 'orange peel' surface. All have excellent wear properties, chemical resistance, and are impervious to grease, oil and water.

	Subfloor	Jointless/Size	Thickness	Manufacturers
Polyurethane resin	Clean, dry, rigid surfaces, ie concrete screeds, granolithic, plywood (as an underlay) blockboard, etc	Jointless	$\frac{1}{16}$″ – $\frac{1}{8}$″	E. J. Elgood Ltd Evode Ltd Limmer & Trinidad Co Ltd Roflex Ltd Quentsplass Ltd Tercol Ltd Torginol Ltd Vigers, Stevens & Adams Ltd
Polyester	similar to above	Jointless	$\frac{3}{32}$″ – $\frac{3}{16}$″	
Epoxy	similar to above	Jointless	$\frac{1}{16}$″ – $\frac{3}{8}$″	

Rubber

SOLID RUBBER; NATURAL OR SYNTHETIC INLAID

Sheet or tiles formed from natural rubber, or from combination of natural or synthetic rubber. A fabric backing is sometimes incorporated for dimensional stability. The characteristics are a quiet resilient floor, easy to keep clean, and with good wearing properties. High resistance to water. Natural rubber has a low resistance to oils, but this can be overcome with the synthetic types such as 'Butile'. The 'rubber' smell which is offensive to some people is said to wear off quickly after laying. Colours: Wide variety of plain, mottled, inlaid pattern and textures. Special ribbed rubber floors pioneered by Pirelli, now generally available from other British manufacturers. Can be laid loose or set into cement screeds.

	Subfloor	Jointless/Size	Thickness	Manufacturers
Solid rubber	Concrete with screed. Timber with hardboard or plywood underlay	Sheet 3′ 4′6″ 6′ (up to 80′ rolls)	$\frac{1}{8}$″ – $\frac{1}{4}$″ up to $\frac{1}{2}$″ available)	Amtico Flooring Ltd Armstrong Cork Co Ltd Harefield Rubber Co Ltd Morris Rubber Industries Ltd Runnymede Rubber Co Ltd
Natural or synthetic inlaid BS1711		Tiles 9″ × 9″ 12″ × 12″ 18″ × 18″ 36″ × 36″ 48″ × 48″	$\frac{1}{8}$″ $\frac{3}{16}$″ $\frac{1}{4}$″	*Ribbed* Pirelli Ltd Redfern Polymers Ltd Runnymede Rubber Co Ltd

Ribbed rubber sheet floor by Redfern Polymers Ltd

SPONGE OR FOAM BACKED

Rubber floors are also made with a thin surface of solid rubber for wear, and a backing of sponge rubber to add greater resilience. The wearing surface is usually from $\frac{1}{16}''$ thick up to $\frac{1}{8}''$, whilst the sponge varies from $\frac{1}{8}''$ to $\frac{3}{8}''$.

Subfloor	Jointless/Size	Thickness	Manufacturers
Concrete with screed. Timber with hardboard or plywood underlay	Sheet 36″ wide to 72″ wide	4.5 mm	Bulgomme-Silence Flooring Co Ltd Harefield Rubber Co Ltd Runnymede Rubber Co Ltd St Albans Rubber Co Ltd

RUBBER CORK

A mixture of rubber and cork granules. It is hardwearing (as other rubber floorings) and is not resistant to oils, but synthetic rubber can be used to overcome this. Quiet, resilient, warm and non-slip. Colours: Limited range of plain and mottled, darkish.

Subfloor	Jointless/Size	Thickness	Manufacturers
Concrete with screed. Timber with hardboard or plywood underlay	Tiles 12″ × 12″ 2′ × 2′ 2′ × 2′6″	$\frac{1}{8}''$ $\frac{3}{16}''$ $\frac{1}{4}''$	British Technical Cork Products Ltd C & I Suppliers Ltd Runnymede Rubber Co Ltd

MATTING

An absorbent alternative to ribbed rubber mats for entrances and surfaces when very hard, dirty wear can be expected. Usually set into a $\frac{3}{4}''$ mat sinking. The construction is by a series of fibre surfaced reinforced rubber strips with intermediate extruded aluminium alloy spacers. Various patterns available. Also mats for external use.

Subfloor	Jointless/Size	Thickness	Manufacturer
Any subfloor	Lengths 12″, 15″, 18″ – 72″ in 3″ increments Widths 9″, 10″, 11″ – 24″ in 1″ increments	$\frac{11}{16}''$	Nuway Manufacturing Co Ltd

Stone

Sandstone	Granite	Limestone
York Stone	Scottish Grey	Portland Stone
Darley Dale	Alps Grey	Hopton Wood
Forest of Dean Blue	Cornish	Ashburton Stone
		Kotah Stone

All very hard wearing stone, which if carefully laid will give almost indefinite life.

Great variety of natural colours from light stone to dark grey. Cast stone slabs are formed using stone chippings in a mix with cement. Such tiles have most of the characteristics of natural stone.

	Subfloor	Jointless/Size	Thickness	Manufacturers
Sandstone	Site concrete with cement bed (externally hardcore and sand or weak cement bed)		2″	
Granite		2′ × 1′6″ – 4′ × 2′6″	1½″	J. Joslin & Co Ltd Nine Elms Stone
Limestone		12″ 15″ 18″ 24″ square tiles	⅝″ – 1½″	
		Slabs 24″ × 60″	2″	

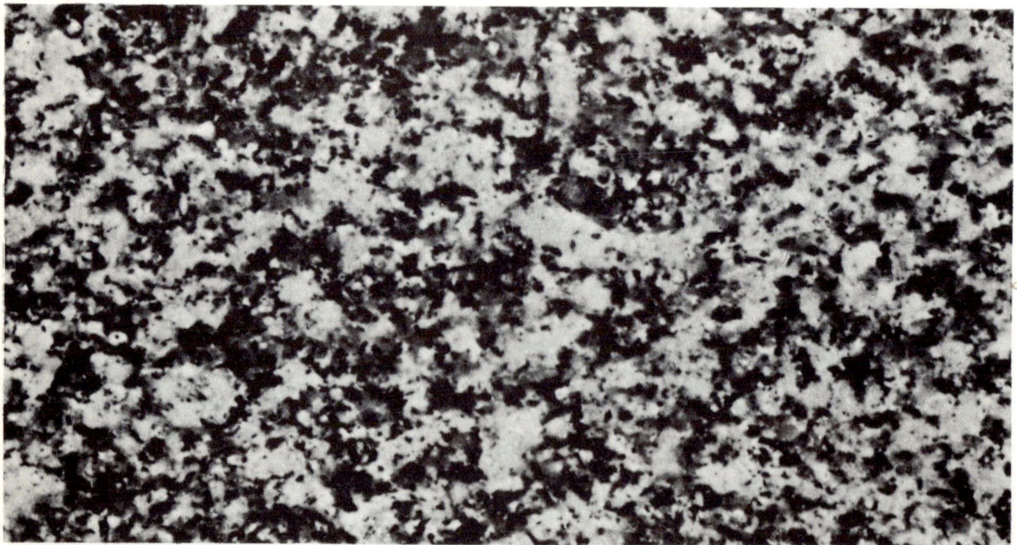

Cornish granite

MARBLE

A centuries-old material which will last for centuries – cold and hard but an infinite variety of colour from white Sicilian to yellow travertine to greens, browns, greys and blacks. Marble is imported from many continental countries and from Ireland. It is expensive and is a superb material to contrast with the warmth of carpet.

Subfloor	Jointless/Size	Thickness	Manufacturers
Site concrete with cement bed (externally hardcore and sand or weak cement bed)	12″ × 12″ 14″ × 14″ 15″ × 15″ 18″ × 18″	⅝″ – ¾″	J. Joslin & Co Ltd A. Elder Reed & Co Ltd Anselm Odling (Construction) Ltd
	up to slabs of 5′0″ × 3′0″	1 3/16″	

QUARTZITE

Quartzite has a riven finish and is non-slip wet or dry. Very hard wearing. Will not chip or crack and is resistant to oils, grease, acids and frost. External or internal use. Can be used for stair treads and risers. Colours vary from silver grey to soft yellows, gold and olive. Alta quartzite has a grey green appearance.

Subfloor	Jointless/Size	Thickness	Manufacturer
Site concrete with cement bed (externally hardcore and sand or weak cement bed)	6″ × 6″ 7″ × 7″ 8″ × 8″ 9″ × 9″ – 24″ × 24″ Other sizes available, eg	$\frac{3}{8}$″ – $\frac{13}{16}$″	Altazite Ltd
	1′11″ × 11$\frac{3}{8}$″ 1′3″ × 7$\frac{3}{8}$″ 1′7″ × 9$\frac{3}{8}$″	$\frac{3}{8}$″ average	

SLATE

Exceptionally hard wearing and good looking. Varies in colour from the green Westmorland slate to dark blue Welsh slate. Can be given a deep gloss appearance if cleaned occasionally with milk. Sawn or polished finishes tend to be slippery when wet. Riven finishes non-slip.

Subfloor	Jointless/Size	Thickness	Manufacturers
Site concrete with cement bed (externally hardcore and sand or weak cement bed)	Welsh 1′6″ × 1′6″ or to random sizes Any sizes can be cut up to 3 sq ft	$\frac{1}{4}$″ – 2″	Bow Slate & Enamel Co Ltd Dinorwic Slate Quarries Co Ltd J. Joslin & Co Ltd Lakeland Green Slate & Stone Co Ltd Nine Elms Stone
	Westmorland Up to 6′ × 3′ × $\frac{3}{4}$″ 2′ × 1′6″ × $\frac{3}{8}$″	$\frac{3}{8}$″ – $\frac{3}{4}$″	The Oakley Slate Quarries Ltd Westmorland Green Slate Quarries Ltd

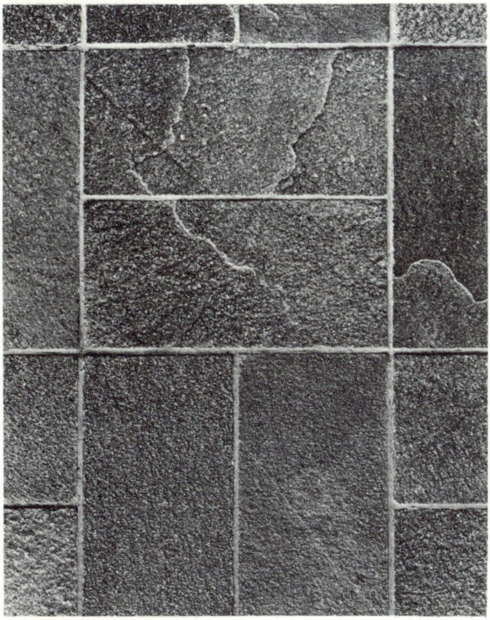

Quartzite stone flooring by Altazite Ltd

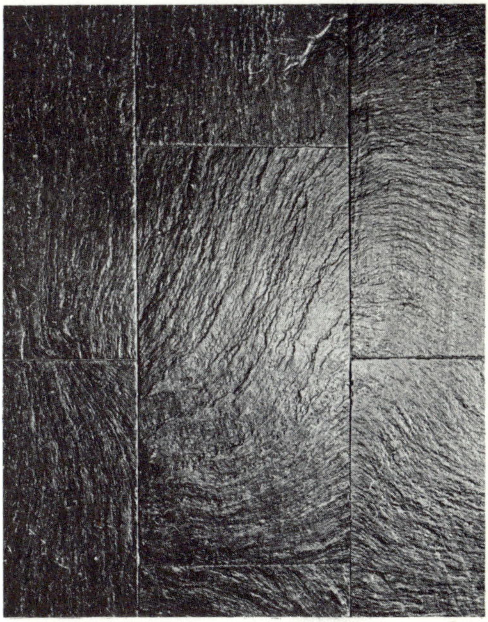

'Wincilate' riven slate floor by The Bow Slate & Enamel Co Ltd

Timber

SOFTWOOD

Softwood boarding is used for the majority of timber suspended floors and is not suitable as a floor finish for areas of heavy use. Softwood block gives a warm rich quality and will wear well for a long life.

	Subfloor	Jointless/Size	Thickness	Manufacturers
Board	Concrete with battens or timber joists	6″ – 7″ wide	1″ – 1½″	Acme Flooring & Paving Co (1904) Ltd J. A. Hewetson & Co Ltd
Block	Cement and sand screed	9″ × 3″ 12″ × 3″	1½″ – ¾″	Vigers, Stevens & Adams Ltd
Strip		3″ – 4″ wide up to 12′ long	1″ – 1½″	

HARDWOOD

Hardwood boards are not much used, although some oak floors have been installed which wear well and look excellent. Hardwood is more traditionally used for block and strip and favourite timbers are oak, maple, sapele, afrormosia and teak, but many others are available. Beechwood strip is manufactured in tongued and grooved sections in which two widths are factory stuck to assist speed of laying in long lengths. Excellent floors for rooms where there is not intensive use. Some hardwood floors can be used for underfloor heating, but specialist advice is necessary. A well maintained hardwood floor improves with age. Colours: Rich variety in natural timbers from light to red or dark browns.

	Subfloor	Jointless/Size	Thickness	Manufacturers
Board	Concrete with battens or timber joists	4″ – 9″	1″ – 1½″	Acme Flooring & Paving Co (1904) Ltd J. A. Hewetson & Co Ltd
Block	Cement and sand screed	12″ × 3″ 9″ × 3″	¾″ – 1½″	Hollis Brothers Ltd Junckers (London) Ltd (strip only)
Strip		2″ – 4″	½″ – 1½″	Vigers, Stevens & Adams Ltd James Webster & Bros Ltd

BLOCK PARQUET/MOSAIC/ENDGRAIN

Block parquets and mosaics are made of solid blocks formed together into panels which are stuck to a rigid subfloor. Mosaics are generally basket pattern, parquets either basket or herringbone. Can be backed with foils or felt. Endgrain blocks are cut from 1″ × 1″ timber to expose the end grain for durability. A traditional high quality floor. Colours: Rich variety in natural timbers from light to red or dark browns.

	Subfloor	Jointless/Size	Thickness	Manufacturers
Parquet	Plywood or other sheet floor	2″ – 4″ wide × 9″	¼″ – ⅜″	Acme Flooring & Paving Co (1904) Ltd
Mosaics	Concrete and screed/asphalt	12″ × 12″ – 18″ × 18″	⅜″ – ⅝″	Contiwood Ltd J. A. Hewetson & Co Ltd
Endgrain	Parquet is laid on a timber subfloor	1″ × 1″ in 12″ × 12″ squares	½″ – 1″	Hollis Bros Ltd The Marley Tile Co Ltd Vigers, Stevens & Adams Ltd Winsorflor Ltd

Softwood block flooring

Teak hardwood block by
J. A. Hewetson & Co Ltd

Wood block mosaic

PLYWOOD PARQUET/CHIPBOARD

Plywood parquet is similar in appearance to block parquet but is formed of plywood to give herringbone or basket weave or strip patterns. It should be remembered that it will only be as good as its top thickness of ply, and is unsuitable for areas of heavy wear. It can be obtained in a variety of finishes from oak to teak, and felt backed tiles are available. Chipboard is now available with a suitable wearing finish, but is useful also as an underfloor to carpet or other tiles – obtainable with polyurethane backing for greater insulation. Special flooring grade chipboard is available with tongued and grooved edges.

	Subfloor	Jointless/Size	Thickness	Manufacturers
Plywood/ parquet	Any rigid dry flat subfloor. Joists must have a sheet ply underlay	9″ × 9″ – 36″ × 36″	⅛″	Contiwood Ltd Par-K-Ply
Chipboard	On timber joists or level concrete/ dry	8′ × 4′ 16″ × 4′	½″ – ¾″	
		(4′ × 2′ with polyurethane foam backing)	1⅜″	

COMPOSITION BLOCK

A composition block formed of cement and wood and linseed oil with keyed underside – set in ½″ cement screed. It is warm, resilient and hard wearing. Can be used with underfloor heating up to 120°F (49°C) surface temperature. Fittings available for skirtings, coves and stairs. Colours: Natural wood and choice of other colours. Vinyl bonded blocks available give wider variety of colours and may be used in areas intermittently wet.

Subfloor	Jointless/Size	Thickness	Manufacturer
Site concrete or preparatory screed	6 3/16″ × 2 1/16″	⅜″ ⅝″	Granwood-Stonewood Ltd
	7½″ × 2½″	½″	

Composition block flooring by Granwood-Stonewood Ltd